PRIZE ESSAYS

ON

ROADS AND ROAD MAKING.

1870.

BOSTON:
WRIGHT & POTTER, STATE PRINTERS,
79 MILK STREET (CORNER OF FEDERAL).
1870.

PRIZE ESSAYS

ON

ROADS AND ROAD MAKING.

1870.

BOSTON:
WRIGHT & POTTER, STATE PRINTERS,
79 MILK STREET (CORNER OF FEDERAL).
1870.

REPORT.

To the Senate and House of Representatives of the Commonwealth of Massachusetts.

In his Inaugural Address to the legislature of 1869, His Excellency Governor Claflin used the following words :—

"Closely connected with our agriculture and other prominent interests is the system of public highways. Few things are of greater importance to a community, or a surer test of civilization, than good roads. Those of our citizens who have visited Europe are unanimous in the opinion that our public roads are far inferior to those of other countries, where the means of easy and safe communication are better appreciated. The science of road-making is, apparently, not well understood; or, if it is, the present modes of superintending the construction and repair of roads are so defective, that the public suffers to an extent of which few of us are aware. It may be found, upon investigating the cause of our miserably poor and ill-constructed roads, that the laws relating to this subject need revision, so as to give more uniformity in the construction and the repair of our highways. It is evident, also, that the science of road-making should have a prominent place in the course of applied mathematics at the Massachusetts Agricultural College."

This suggestion was embodied by the legislature in the form of a Resolve, offering prizes for an Essay or Essays upon Roads, in the following words :—

Resolve concerning the construction and repair of roads.

Resolved, That there be allowed and paid out of the treasury of the Commonwealth the sum of four hundred dollars, to be expended under the direction of the board of agriculture, in the payment of one or more premiums for the best treatise or treatises, containing not more than two hundred pages duodecimo, respectively, upon the science of road-making and the best methods of superintending the construction and repair of public roads in this Commonwealth; and that said board are also authorized to cause to be printed, for the use of the next legislature, three thousand copies of the treatise receiving the highest premium under this resolve, if they shall deem such publication expedient. [*Approved, June* 12, 1869.

The subject was referred by the Board to a Committee consisting of Messrs. James F. C. Hyde, of Newton, Avery P. Slade, of Somerset, and Newton S. Hubbard, of Brimfield, and this Committee was authorized to offer, on behalf of the Board, one or more prizes as, on mature consideration, it should think best. The Committee issued the following circular:—

ROADS, AND ROAD MAKING.

The undersigned, a Committee of the State Board of Agriculture, would respectfully call your attention to a Resolve, adopted by the Massachusetts legislature at its last session, and approved June 12th, 1869, "concerning the construction and repair of roads;" a copy of which is hereto annexed.

The Board of Agriculture offer $200 for the best, $125 for the second best, and $75 for the third best treatise on the subject of road making and repair, as provided for by the Resolve; the same to be sent to Charles L. Flint, the Secretary of the Board of Agriculture, State House, Boston, on or before January 28th, 1870, for examination. All manuscripts, whether receiving an award or otherwise, to be at the disposal of the Board. As the time is short, it is earnestly desired that the subject may receive immediate and thorough attention, as it is one of very great importance to the whole State.

JAMES F. C. HYDE,
AVERY P. SLADE,
NEWTON S. HUBBARD,
Committee.

DECEMBER 13, 1869.

A wide-spread interest was awakened in the subject. Nearly thirty essays were submitted for the inspection of the Committee, some of them by writers residing in other States. These essays differed widely in their character. Some of them were scientific and elaborate; others meagre in extent though often containing suggestive and useful hints. No one appeared to the Committee to embody all that was desirable, or to form in itself a full and comprehensive treatise adapted to the wants of both cities and towns, or large centres of population and sparsely inhabited districts.

The general principles of road making may be the same, but it is evident that they cannot be applied with equal completeness under all circumstances. A city street, or a road in the neighborhood of large and wealthy centres of population is one thing, and a country road, little frequented, is another. A sandy road on Cape Cod, where both clay and gravel are inaccessible, and a rocky mountain road in Western Massachusetts, would require somewhat different treatment.

Nothing certainly need be said of the importance of good roads. No community can afford to be without them. The wear and tear of vehicles and horse-flesh and muscle on a bad road is enormous; and it constitutes an indirect tax which probably amounts to very nearly as much as the direct tax upon the people. Strange as it may seem, public sentiment needs to be educated up to a full appreciation and realization of the difference between good and bad roads. Nor can it be questioned that the system of road making, as it has been adopted and applied hitherto in most of our towns, is defective and bungling. True economy would require a more liberal original outlay, if the public comfort and convenience did not, for the item of cost for repairs to follow would be materially and permanently lessened. Besides, the same power will move a much larger load, and do it with greater ease, with less wear and tear, on a solid, properly-constructed road, than on one of an opposite character, to say nothing of a great saving of time by the greater degree of speed attained.

The first prize was assigned by the Committee to CLEMENS HERSCHEL, of Boston; the second to Prof. SAMUEL F. MILLER, of Amherst; the third to HENRY ONION, of Dedham.

The awards were made, after very careful study and consideration, with reference to the wants of all sections of the State, and though neither of the following Essays would be complete in itself, it is believed, that, taken together, as they should be, they present a comprehensive view of the subject, and that they will be found both useful and instructive, as furnishing suggestions worthy to form the basis of legislative action.

CHARLES L. FLINT,

By order of the Committee.

A FIRST PRIZE TREATISE

ON

THE SCIENCE OF ROAD MAKING.

By Clemens Herschel, of Boston,

civil engineer, member of the american society of civil engineers.

Introduction.

This treatise was written in answer to the printed circular of a Committee of the Board of Agriculture, calling for "treatises upon the science of road-making, and the best methods of superintending the construction and repair of public roads in this Commonwealth."

This circular was issued about the middle of December, and as the time for writing and sending in the called-for essays was limited to January 28, the writer has thought it best, no specific character being prescribed for the treatises, to attempt to write one suitable to be so called from the stand-point of the *public*, rather than from that of the civil engineer, and, giving results rather than the methods of arriving at them, to be as concise as possible.

The Science of Road Making.

Starting, then, with the first of the two subjects mentioned in the circular,—the science of road making, we can divide this into three periods: (1) laying out a road; (2) making the road-bed, which includes all earthworks, cutting and filling, culverts, drains, bridges, even tunnels, &c.; and (3) the making of the road surface; to which, not improperly, might be added, (4) keeping the road in repair.

Laying out a Road.

The considerations which determine the best location of a road, are those arising from the nature of the travel it is proposed to accommodate; that is, from the admissible grades, radii of curves, &c. Given two points it is desired to connect, with no intermediate point where the road is to touch, that route is the best which will cost least to build and maintain, the grades and curves being kept within bounds; and to find this location constitutes the whole problem of the engineer.

The Romans built all of their roads in perfectly straight lines, up hill and down, at a very great expense, as being absolutely the shortest distance between two points. At a later period in history, it was argued that a road *must* be winding to be agreeable, and many were so built only for this reason. The modern road-builder or engineer in general, ignores any such considerations, and has for his aim only to achieve the most, at the least present and future expense.

As regards curves in roads in a hilly or mountainous district, we have then the rules never to make a smaller radius than 20 feet, and that only in extraordinary cases. On roads where long logging or other wagons may be expected, the smallest radius ought to be 50 or 60 feet; and, in general, 40–45 feet is none too much.

A rule sometimes followed in constructing mountain roads, is, where the inclination is 1 or 2 in a hundred,* heavy teams require 40′ and light ones 30′ radius; with a grade of 2 or 3 in a hundred, heavy teams require 65′ and light ones 50′ radius. Where a reverse curve (shaped like the letter S) occurs, there should be a straight piece connecting the two curves, (Fig. 1.) On the contrary, where the two curves to be connected are concave in the same direction, the connecting link should be curved also, and not straight, (Fig. 2.) On the length of the curves the grade should be made easier than on the parts of the road immediately adjoining.

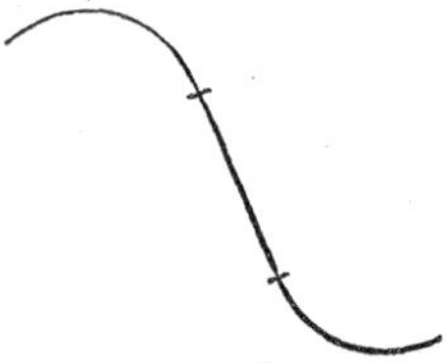

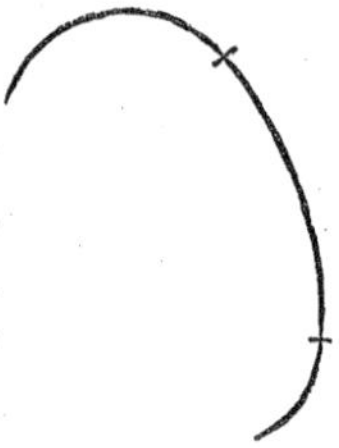

FIG. 1.

FIG. 2.

As regards grades, to start with mountain paths, we find pedestrians able to walk up an inclination of 100 in 120;

* In describing grades, the first figure gives the vertical height which is ascended in a horizontal distance given by the second figure. Both figures must of course be taken to refer to the same unit of length, thus: 100 feet

mules, ponies, &c., 100 in 173. For roads, Telford's rule was, that for horses attached to ordinary vehicles to trot up a hill rising 3 in one hundred, was equal to walking up one of a 5 in a hundred grade.

Experiments have shown that,—

1. On a road falling 2 in a hundred, vehicles would run down of themselves.

2. On the same kind of road, but having an inclination of 4 in a hundred, light vehicles had to be held back lightly, loaded ones, with considerable force.

3. On a road having a fall of $5\frac{1}{2}$ in a hundred, light vehicles had to be held back with considerable force, or if a brake was applied they had to be pulled, whereas heavy or loaded vehicles had to be braked to keep the horses from being speedily exhausted.

On inclinations steeper than 5 in a hundred, the rain-water running down the road is apt to do some damage to the road surface.

The regulations of different countries having a long experience in road building, such as France, Prussia, Baden, &c., vary somewhat, but the following is the general result.

In treating of roads, it often renders the subject much clearer, to divide them into three classes: first, second and third class roads, or, as we might also say, state, county and town roads. Accepting this nomenclature, we have this: for first class or state roads, the greatest inclination should not exceed 3–5 in a hundred; second class or county roads, 5–7 in a hundred; third class or town roads, 7–10 in a hundred. A road rising 10 in a hundred is not supposed ever to have any heavy teams upon it. In ascending a hill it is well and proper to decrease the grade as the top is reached, and in the same measure as the horses get tired. Thus, if a first class road starts up hill with a grade of $4\frac{1}{2}$ per hundred, it should gradually diminish to 4 and $3\frac{1}{2}$ in a hundred, and end near the top with a grade of 3 in a hundred. If a grade of 4 or 5 in a hundred must needs be kept up for some distance, then it is well to have resting places 40 or 50 feet

in 120 feet, 100 inches in 120 inches, or 100 miles in 120 miles, all express the same inclination to a level plane, and are more general in their application than the ways of expressing grades in so many inches to the foot, or feet in one mile, &c., &c.

long, having a grade of only $1\frac{1}{2}$ or 2 in a hundred, in the line of the road at proper intervals. An expedient adopted by Telford, the eminent English engineer, in order to avoid making a piece of road a mile long, on a less grade than 5 in a hundred, on account of the increased cost this would have occasioned, and yet not have this part of the road too much more tiresome for the horses than the rest, was to make the road-surface on this mile of a much better quality than on the remainder; the additional cost required for the improved road-bed amounting to only about one-half of what it would have cost to reduce the grade to say 4 in a hundred, as will be again referred to under the head of trackways. In sharp curves the grade should be only 1 or 2 in a hundred or level.

The following table gives the effects of various grades on the amount a horse can pull, and is based on calling the load a horse will pull on a level, one:—

Then, on a grade of	1:100,	a horse can pull		. .	0.90
"	"	1: 50,	"	" . .	0.81
"	"	1: 44,	"	" . .	0.75
"	"	1: 40,	"	" . .	0.72
"	"	1: 30,	"	" . .	0.64
"	"	1: 26,	"	" . .	0.54
"	"	1: 24,	"	" . .	0.50
"	"	1: 20,	"	" . .	0.40
"	"	1: 10,	"	" . .	0.25

To determine whether it is most advisable to go over or around a hill, all other considerations being equal, we have this rule: Call the difference between the distance around on a level and that over the hill, *d*, the distance around being taken as the greatest, and call *h*, the height of the hill.

Then in case of a first class road, we go around when *d* is less than 16 *h*.

And in case of a second class road, we go around when *d* is less than 10 *h*.

When the height of a necessary embankment gets to be more than 60 or 65 feet, a bridge or viaduct will be found cheaper, and the same measure, 60 feet, applies in case of tunnels, they being cheaper at that depth than open cuttings.

Under the head of laying out roads, something should be said of their width. Speaking only of such roads as are not apt to turn into streets from their proximity to towns and cities, it is well not to make them too broad, for the less the width, the less the cost of construction and maintenance, and a good 23 feet road is much better than a poor one 40 or more feet wide. Each rod ($16\frac{1}{2}$ feet) in width adds two acres per mile to the road. An agreeable form of road is to have on each, or on one side of the same a strip, 5 or 6 feet wide, sodded, and then a sidewalk equal in width to one-eighth the width of the roadway. The intervening strip above mentioned, is planted with trees and at intervals of 200–250 feet furnishes storage places, 30 or 40 long, for the materials used in the road repairs. The width of first, second and third class roadways may be given as 26, $18\frac{1}{2}$ and 13 feet, with a tendency during the last ten years to have none, except in the vicinity of cities, wider than 24 feet, and the rest correspondingly narrower. In view of the changes constantly going on in this country in the value and settlement of land, it would probably be well always to *lay out* a road 50 or 60 feet wide, but to *build* the road proper of the widths above indicated.

With all these rules and data in mind, the real work of actually laying out the road on the ground and on a map is next in order, and this comes so entirely within the province of the civil engineer, and is a matter requiring so much explanation and study, that it cannot well be introduced within the limits of this treatise. It is in this part of the work that a little skill and labor well spent may be productive of very great saving in the cost of the whole work and it should not be left to the inexperienced or unskilful.*

* Gillispie, in his treatise on "Roads and Railroads," gives two forcible instances of the amount those roads which might properly be called *chance* roads, can be improved by a road-maker of skill and understanding. An old road in Anglesea, England, rose and fell, between its two extremities, twenty-four miles apart, a total perpendicular amount of 3,540 feet; while a new road, laid out by Telford between the same points, rose and fell only 2,257 feet; so that 1,283 feet of perpendicular height is now done away with, which every horse passing over the road had previously been obliged to ascend and descend with its load. The new road is besides two miles shorter. The other case is that of a plank-road built in the State of New York, between the villages of Cazenovia and Chittenango. Both these villages are situated on Chittenango Creek, the former being eight hundred feet higher than the latter. The most level common road between these

Making the Road-bed.

Under this head are included, earthworks, drains, culverts, bridges, stay walls, &c., &c., all matters requiring a special kind of skill to construct properly. The writer believes it impracticable to write a book which shall at once be interesting to and therefore valued by the public, and of value to the professional man, and thinks an attempt so to do results always in a failure in both directions. True to the determination expressed in the introduction, he proposes, therefore, to treat under this head mainly with those parts of the subject in which the public at large is most interested, for example, the data for the cost of earthworks, general information relating to drainage, bridges, &c.

Earthworks.

The basis of all values is the daily wages of a common unskilled laborer, and in the data given below, this figure, whatever it is from time to time and in various places, must be taken as unity, or the standard measure.

The cost of earthworks may be divided into three parts—(1) cost of loosening the earth, (2) cost of transport, and (3) cost of forming the transported earth into the desired shape. The cost of the first part depends materially on the kind of earth to be handled. The cost of the second, mainly on the distance the earth is to be moved.

We find by experience, that in digging and loading or throwing 5–10 feet horizontally with a shovel, we obtain for different materials the results of the following table:—

villages, rose, however more than 1,200 feet in going from Chittenango to Cazenovia, and rises more than four hundred feet in going from Cazenovia to Chittenango, in spite of this latter place being eight hundred feet lower. That is, it rises four hundred feet where there should be a continual descent. The line of the plank-road laid out by George Geddes, civil engineer, ascends only the necessary eight hundred feet in one direction, and has no ascents in the other, with two or three trifling exceptions of a few feet in all, admitted in order to save expense. The scenes of similar possible improvements are scattered all over this and the rest of the States; and these facts are still more or equally to be borne in mind in laying out new roads, where the ounce of prevention may take the place of the pound of cure.

Number.	KINDS OF EARTH.	In Parts of a Laborer's Day's Wages.			Amount to be added for keeping and repairing tools.
		Cost per cubic yard to loosen.	Cost per cubic yard to load in wheelbarrows.	Cost per cubic yard to load in carts or wagons.	
1	Loose earths, which are loam, sand, &c., inclusive of loading,	$\frac{1}{7}-\frac{1}{9}$	–	–	$\frac{1}{20}$
2	Heavier earths, such as sticky clay, which does not readily leave the shovel, &c.,	$\frac{1}{6}-\frac{1}{7}$	$\frac{1}{16}$	$\frac{1}{12}$	$\frac{1}{20}$
3	Earths which must be loosened with a pick before they may be shovelled,	$\frac{1}{4\cdot5}-\frac{1}{5}$	$\frac{1}{16}$	$\frac{1}{12}$	$\frac{1}{15}$
4	Solid banks of gravel or clay, earths containing boulders, &c., in which one man only loosens as much as another man shovels,	$\frac{1}{4}-\frac{1}{4\cdot5}$	$\frac{1}{16}$	$\frac{1}{12}$	$\frac{1}{12}$
5	Same material, worst kind, brick and mortar heaps, earth full of roots, &c., in which it takes two men to loosen what one man shovels,	$\frac{1}{3}-\frac{1}{3\cdot5}$	$\frac{1}{12}$	$\frac{1}{8}$	$\frac{1}{10}$
6	To break up stone which is in layers or seams, requiring the use of the crowbar only, but no blasting,	$\frac{1}{2\cdot5}-\frac{1}{3}$	$\frac{1}{12}$	$\frac{1}{8}$	$\frac{1}{10}$
7	Blasting rocks in an open cut, according to the hardness of the rock, to the position of the seams, &c.,*	$\frac{1}{2\cdot5}-\frac{1}{1\cdot5}-\frac{1}{1\cdot25}-\frac{1}{0\cdot75}$	$\frac{1}{12}$	$\frac{1}{8}$	$\frac{1}{10}$
8	In forming and shaping embankments,	$\frac{1}{25}$	–	–	$\frac{1}{20}$

* To excavate rock to a given line and level, that is, to trim a cutting may cost double these figures per yard.

Transport of Earth.

Throwing with a shovel.—This is to be done only from 5–12 feet in distance or from 5–6 feet vertically. To throw 5 feet vertically, costs as much as 12 feet horizontally, that is to say, if 30 feet horizontally cost per cubit yard, $\frac{\text{1 day's wages}}{8.4}$ the same distance vertically will cost about $2\frac{1}{2}$ times as much, or more exactly, $\frac{\text{1 day's wages}}{3.5}$ whence is seen the economy of using windlasses, &c., instead of "stages,"* in shovelling earth vertically. The table gives the cost of shovelling earth certain distances, expressed in the number of cubic yards a laborer's day's wages will pay for.

Distance of Throw, in Feet.	Vertical or Horizontal.	Whether done at one operat'n, or by means of so-called "stages."	Number of cubic yards which can be transported at the cost of one laborer's day's wages.	Remarks.
0–10, . .	Horizontally,	No "stages."	23.5	
10–20, . .	"	1 stage.	12.6	Wheelbarrow cheaper.
20–30, . .	"	2 stages.	8.4	
0–5, . .	Vertically,	No stages.	14.1	
5–10, . .	"	1 stage.	8.8	

Wheelbarrows.—The usual distance of transport suitable for the use of wheelbarrows is 100–200 feet. In exceptional cases it may be more, but perhaps never above 500 feet, and then only for moderate quantities. In going up hill, the greatest inclination is to be not more than 1 in 10, and a man can push only $\frac{2}{3}$ as much on this inclination as on a level. 3 feet vertical transport costs as much as 90–100′ horizontally. Whenever possible, planks should be laid for the wheel-barrows to run on. The best timber for this purpose is beech wood and the cost of keeping such planks is only about $\frac{1}{40}$ or $\frac{1}{50}$ per cent. of the cost of transport per cubic yard.

* By a "stage," is meant the operation of one shoveller lifting and throwing what another has thrown in front of him.

DISTANCE OF TRANSPORT, IN FEET.	Number of trips per day of ten hours, made with one man at barrow, and one to load.	Contents of wheelbarrow load in cubic feet.	Number of cubic yards which can be transported at the cost of one laborer's day's wages.
10-20, ...	120	2½	23.5
20-50, ...	110	2½	16.9
50-70, ...	100	2½	14.4
70-100, ...	98	2½	13.8
100-150, ...	96	2½	13.3
150-200, ...	94	2½	12.8
200-250, ...	92	2½	12.4
250-300, ...	90	2½	12.0
300-350, ...	88	2½	11.6
350-400, ...	86	2½	11.2
400-450, ...	84	2½	10.9
450-500, ...	82	2½	10.5
500-550, ...	80	2½	10.2

PATENT PORTABLE RAILROAD AND HAND CARS.

These have lately been introduced in this country and appear to be coming into general use and favor. The company owning this improvement, as it seems to have a right to be called, claim, that by means of their track and cars, which can be used everywhere that a wheel-barrow or a horse-cart can go, and in a great many places where these vehicles cannot go, they effect a very large saving, as much in some cases as $\frac{5}{6}$ of the cost by the other means of transport. There are no data published as yet to make tables from similar to the foregoing; from the company's pamphlet, however, one given case which occurred on Staten Island in 1867, may be analyzed and tabulated as follows:—

Distance of transport, in feet, ... 550

Number of trips per day of ten hours, with one man at two cars, and two to load, ... 150
Contents of car, in cubic feet, ... 11.34
Number of cubic yards which can be transported at the cost of one laborer's day's wages, ... 60

One-Horse Carts.

The table for this kind of transport may be stated about as follows. 1 foot vertical costs as much as 14 horizontal.

Distance of Transport, in Feet.	Number of trips made per day of ten hours, assuming only four minutes to load, dump, &c., per trip.	Contents of cart load in cubic feet.	Number of cubic yards which can be transported at the cost of a laborer's day's wages.
300, ...	86	8	17.1
500, ...	67	8	13.6
1,000, ...	43	8	8.6
1,500, ...	31	8	6.4
2,000, ...	25	8	5.0
2,500, ...	21	8	4.3
3,000, ...	18	8	3.6

Ox-cart transport is 10 or 12 per cent. cheaper than the above, but takes more time.

Other methods of transport, such as horses or engines on temporary tracks, would hardly ever be applied to road-building, but belong more properly under the head of railroad construction.

Shrinkage.

In calculating the cost of earthworks, the so-called shrinkage of earth must not be overlooked. Earth occupies on the average $\frac{1}{10}$ less space in embankment than it did in its natural state, 100 cubic yards, shrinking into 90. Rock on the contrary, occupies more space when broken, its bulk increasing by about one-half.

The shrinkage of gravelly earth and sand may be taken at 8, of clay 10, loam 12, surface soil 15, and of "puddled" clay 25 per cent. The increase of bulk of rock is 40 to 60 per cent.

To make use of all these data in calculating the probable cost of a piece of road, there is of course still wanting the equally essential factor which gives the number of cubic yards to be dug and moved and the distance of transport. These are got from the plan, profile and cross-sections of the proposed work, an engineer's knowledge being requisite to make the necessary drawings and calculations.

Drains and Culverts.

The drainage of roads is of two kinds, surface and sub-drainage. The first provides for a speedy removal of the rain-fall on the surface of the road and the cutting and embankments on which it is carried; the second, for the removal of that part of the rain-fall which nevertheless does penetrate into the body of the road-covering. With a perfect sub-drainage the winter's frost, having no water to act upon within the body of the road, is robbed of its great power to destroy the same, and it also prevents the road-surface from becoming soaked and thence destroyed in the summer. The need of surface drainage is self-evident. This last named is to be provided for at this stage of the building of the road, the sub-drainage being more properly a part of the building of the road-covering or top. For this purpose ditches, one on each side generally, are absolutely necessary, both when the road is on a level with the surrounding country and when it is in a cutting. They may become necessary also in the case of embankments; for example, when an embankment is built across low ground. Where these side ditches cross under the embankment we have a culvert; also whenever any small valley, having a constant or intermittent stream of water, is crossed by such an embankment. It is very bad policy to make such culverts of wood, unless indeed they are so situated as to be constantly under water; the cost of replacing them after the embankment and road has been built over them is disproportionately great. They should be made of stone, or brick; lately, cement drain-pipe, oval or egg-shaped, has been used to advantage in their construction.

All ditches, drains and culverts should have a fall throughout their entire length. Their size will depend on the amount of water they may be expected to carry and this again on the rainfall that may occur on the area which they drain. Extraordinary showers have occurred of 2″ in half an hour but only over a very limited area and 2″ in an hour may be taken as a large allowance. This is the basis of the Central Park drainage calculations and is larger than usually taken, none too large however for safety. As culverts grow larger and wider with the amount of water they are to pass under the road, they develop finally into

Bridges.

Bridge-building is a life's study, taken by itself, and in some of its parts it is not half appreciated and known as yet among the public. Prominent among these is beauty of design and *appropriateness to the situation.* There is perhaps nothing else that will so much improve the appearance and attractiveness of a road as a beautiful bridge. So also in cities we find that a street will of its own accord, seemingly, improve in appearance, when a good and handsome bridge has been erected on its line, the owners and builders of the adjoining buildings taking the bridge for their pattern and model. Nor must it be supposed that a handsome bridge must necessarily cost more than an inappropriate or homely, uncouth structure ; it need never be the case. Very often the chief beauty of a structure lies in the fact of its carrying the most with the least expenditure of material. No one bridge is proper in every situation and herein are many mistakes made. The correct way to build a good bridge, is the same or a similar way to that followed in first-class buildings, namely, to have plans drawn for the same and receive estimates and offers to build according to these plans. It is not well to allow the offices of designer, superintendent and contractor to be united in one person or firm, and is expecting too much from human nature.

Making the Road Surface.

There are two subordinate kinds of road surface, if the term road can properly be applied to them, namely, that of foot and riding paths ; these may be disposed of first, before proceeding to the more important consideration of the road surfaces proper, those used by vehicles of all descriptions.

Footpaths.—For the surface of a footpath little solidity is necessary, except in city sidewalks which are not supposed to be treated of here, but we do need a material that shall become and stay compact soon after it is laid. Coarse sand, screened gravel, stone chips and dust, make good paths; should these materials be too free from any earth or clay, a little of the same may often be added to advantage to act as a binding material. Wherever the ground underneath the surfacing is not porous or likely to remain porous enough to let all the water that may soak through, drain away, a layer of such porous material must be filled up before the top surface is put on. Oyster shells, or large stone chips, gravel stones or pebbles, &c., make a good foundation of this sort. The top covering should have a slope, best in both directions from the centre of the path towards each side of about 1 in 16; the thickness of the foundation course to be 3–5, and that of the top 3–4 inches. *Heavy* rolling will save much time in finishing the whole process; the roller should be used unsparingly and throughout the whole construction of the path, on the foundation, as well as on the top.

Riding-Paths.—From the nature of the travel these are intended to accommodate, their surface is of a peculiar nature. Inasmuch as a horse, in galloping, tends to throw the soil he treads on backwards with his hind feet, the surface must be kept somewhat loose and soft to make riding on it easy and agreeable. This requirement makes it impossible to have any slope on the surface (the loose material would wash away if there were any), and hence we must rely here wholly on sub-drainage, and not attempt any surface drainage. The top is made of coarse sand, *free from clay* or other binding material, laid on two and one-half to three and one-half inches thick, and spread out level. Under this is a solid foundation, about four inches thick, made of coarse gravel and clay, and having a slope of about 1:20, so that the water will run off along its top surface to either side, where it must further be disposed of by drains or ditches. In case of riding paths too wide to be so simply built, the sketch shows the method to be used. The

foundation is made in several slopes, at the lowest parts of which are placed drains, running in the direction of the path, but communicating from time to time with the side ditches or drains. Should, however, the ground underneath be porous enough, the drains may be dispensed with; and if in their stead holes be dug along the lowest lines, marked *a*, *a*, and these filled with large stone, the water will, through them, drain away into the ground.

Roads.—To make a good road surface is a very simple operation after it is only once understood, and, the fundamental principles thereof once comprehended, they can hardly be forgotten. Everything connected with the construction, the use and maintenance of roads, was, in times past, before the invention of railways, the subject of exact observations and experiments, many and varied in their character. Besides this, we have the results of a great number of years of experience in older countries, and there would seem to be little to invent, but much to learn, in this branch of construction. Though less progressive than other branches, there are nevertheless improvements in road-making, especially in road-making machinery and tools; and no treatise on this or any other living subject can be considered complete a very few years after it is written.

Ancient roads were made with a surface as nearly resembling the solid rock as possible. So, in China, roads were made of huge granite blocks laid on immovable foundations. In time these became worn with ruts, especially in the joints or seams of the stones, and the surface generally so smooth that animals could hardly stand, far less trot on it. They are now for the most part deserted, and left to be covered up by land-slides, &c., to one side of the new roads of travel.

The invention of McAdam consisted in having no large stone at all on the roadway, but having it all pounded into small fragments and spread over the road-bed. This has, without fear of efficient contradiction or shadow of doubt, been proved by trial to be a worthless proceeding, though at one time popular, and even now only too often done, either from ignorance or laziness. The separate fragments of stone, having no bond among themselves, are liable to sink into the underlying ground or road-bed, evenly or unevenly as it may chance, more in one place than in another, and thus never come to rest or to an even top surface. Between these two extremes of an ancient Chinese solid rock road and that of McAdam, lies the true principle of road-making, which consists in giving every road two component parts; one,—the foundation,—to be solid, unyielding, porous, and of large material; the other—the top surface—to be made up of lighter material, and to be made to bind compactly and evenly over the rough foundation. This constitutes the whole principle to be followed; and let it be repeated, that to dump the road material directly on the ground, without first preparing a foundation for it, as is so frequently done, is a waste of time, labor and materials, by no possibility resulting in a good road. On this one fundamental idea, which is never abandoned, however, there are a number of variations. Besides these roads, whose characteristic is the foundation they are all built on, we have paved roads, or pavements, of a great many kinds, and roads with trackways, also of various kinds.

Foundation Roads.

The roads of this kind, with macadam for the top surface, are called Telford roads by English writers, from Telford, who first built them in England. The Central Park "gravel roads" belong under this head, gravel taking the place of the macadam of the Telford roads. These foundation roads are of far greater importance than any other kind for State, county or town roads, also for parks and driveways. The top surface of all these roads must have a certain inclination, to cause efficient surface drainage. Various authorities give various rules for the amount of this inclination or side-slope. It would seem just that it should depend on the nature of the top covering, being less for more solid than for looser or softer materials, and also on the grade of the road.

In Baden, one of the smaller German States, but which is worthy to be taken as a model in matters of road-building, and in France, the rise at the centre is given as $\frac{1}{40}$–$\frac{1}{60}$ of the width of the road, according to the nature of the material; that is, inclinations of 1 in 20, and 1 in 30. The rules in Prussia prescribe inclinations of 1 in 24 for roads falling more than 4 in a hundred; 1 in 18 for roads on a grade of between 2 and 4 in a hundred; and 1 in 12 for those on a grade of less than 2 in a hundred. When first built, the centre should be made some four inches too high, to allow for after settling.

Macadam Top.—The cross section of such a road is about as drawn ; 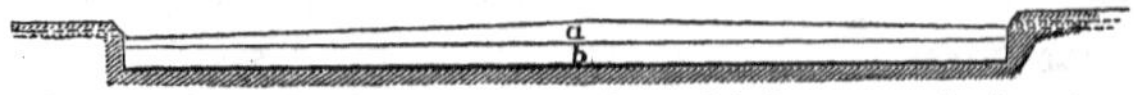 the thickness of the foundation $b=a$, the thickness of the top covering at the centre, and is six, four or five and three and one-half inches in thickness for first, second and third class roads. If the stone for the foundation—for which most anything will do, and that kind should be taken which is cheapest to procure —happens to be got out cheapest in larger pieces than the above dimensions, it will do no harm. This foundation course is sometimes set so as to present an inclination on top, and the cover then put on of a uniform thickness over the whole breadth. This is perhaps best, but is somewhat more expensive. It will do, in nearly all cases, to set the foundation course on a level, or as near so as the stones will allow, and then make the top crowning, by making the covering say three-quarters of an inch or an inch less thick at the edges than in the centre. The stones forming the foundation should not be set in rows, nor ever laid on their flat sides, but set up on edge and made to break joints as much as possible ; that is, set up irregularly. After they are set up, the points that project above the general level may be broken off, and the interstices generally filled with small stone. More or less care and work are necessary in this part of the operation, according to the importance of the road and the depth and character of the material used for the top covering. To roll the road at this stage is to be recommended ; afterwards it becomes a requisite. The point never to be lost sight of, is that this foundation course must remain porous, must be *pervious* to water, so that all rain-water that shall soak through the top covering will find, through it, means of escape to the ground underneath ; thence, according to the nature of the subsoil, it is left either to soak into the ground, or must be further led away by appropriate drains.

Of very great importance is the *material* used for the top or road covering. In the order of their value for macadam, we have.

I. Basalt.
II. Syenite and Granite.
III. Limestones.
IV. Sandstones.

It will be evident, that a very much greater quantity of the soft stones would be required to repair a certain road, than of a harder kind, and on a road lying out of the way of a hard stone quarry or deposit, the question will arise which is cheapest, to pay more for the raw material and get good stock, or pay less and use the worse. There have been some interesting results in places where this matter has been the subject of experiment, continued for a number of years. Thus on a road in Baden which was formerly macadamized with rock costing only fifty cents per cubic yard, it was finally found cheaper, to take harder rock from a distance costing one dollar and seventy-eight cents per cubic yard, the saving being both in less *quantity* of material used and less *labor* required in repairs. Just where the limit is, must be found in each case by long continued experiment, which it is well worth the trouble to make, both to save expense and also to have the best possible road, the harder material making a road better at all times, at the same or less cost. After the right kind has been determined, none other should be mixed with it, and should any inferior piece accidentally or designedly get into the stock to be broken up, it should be picked out and thrown aside. The stone is broken up into macadam, either by hand or machinery. Wherever any considerable quantity of macadam is in present or future demand, a stone-breaker is certainly a saving over hand-labor, though it is difficult to draw the line exactly, where hand-labor or machine labor is cheapest. Probably no town that pretends to keep thirty or forty miles of road in good repair, ought to be without one of these labor-saving machines. Those most in use are made by Blake Bros. of New Haven, Conn., and the following is taken from their circular.

Their machine has been patented in the United States and in several foreign countries, and is now in use in almost every country on the globe. It is simple and compact, and being complete in itself, requires no extraneous support or fixtures.

Description.—Figure 3, is a perspective view of the machine, entire. The frame A, which receives and supports all the other parts, is cast in one piece, with feet to stand upon the floor or on timbers. These feet are provided with holes for bolts, by which it may be fastened down if desired; but this is unnecessary, as its own weight gives it all the stability it requires. B,

B, are fly-wheels on a shaft which has its bearings on the frame, and which between these bearings, is formed into a short crank.

FIGURE 3.

C, is a pully on the same shaft, to receive a belt from a steam-engine or other driver.

Figure 4, shows a side view or elevation of the other parts of the machine in place, as they would be presented to view by removing one side of the frame. The parts of this figure which are shaded by diagonal lines, are sections of those parts of the frame which connect its two sides, and which are supposed to be cut asunder, in order to remove one side and present the other parts to view. The dotted circle D, is a section of the fly-wheel shaft; and the circle E, is a section of the crank. F, is a pitman or connecting rod which connects the crank with the lever G. This lever has its fulcrum on the frame at H. A vertical piece I, stands upon the lever, against the top of which piece the toggles J, J, have their bearings, forming an elbow or tog-

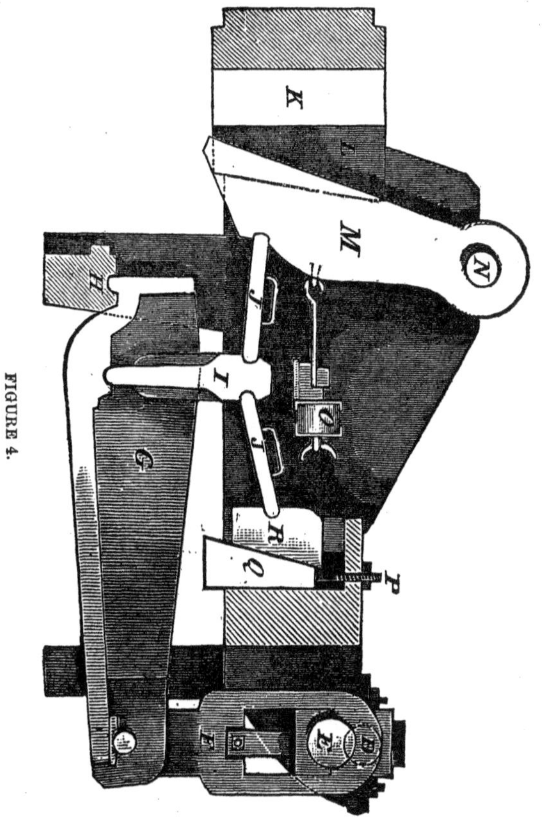

FIGURE 4.

gle-joint. K, is the fixed jaw against which the stones are crushed. This is bedded in zinc against the end of the frame, and held back to its place by cheeks L, that fit in recesses in the interior of the frame on each side. M, is the movable jaw. This is supported by the round bar of iron N, which passes freely through it and forms the pivot upon which it vibrates. O, is a spring of india-rubber, which is compressed by the forward movement of the jaw and aids its return.

Every revolution of the crank causes the lower end of the movable jaw to advance toward the fixed jaw about $\frac{1}{4}$ of an inch and return. Hence if a stone be dropped in between the convergent faces of the jaws, it will be broken by the next succeeding bite ; the resulting fragments will then fall lower down and be broken again, and so on until they are made small enough to pass out at the bottom. The readiness with which the hardest stones yield at once to the influence of this gentle and quiet movement and melt down into small fragments, surprises and astonishes every one who witnesses the operation of the machine.

It will be seen that the distance between the jaws at the bottom, limits the size of the fragments. This distance, and consequently the size of the fragments, may be regulated at pleasure. A variation to the extent of $\frac{5}{8}$ of an inch may be made by turning the screw-nut P, which raises or lowers the wedge Q, and moves the toggle-block R, forward or back. Further variations may be made by substituting for the toggles J, J, or either of them, others that are longer or shorter ; extra toggles of different lengths being furnished for this purpose.

This machine may be made of any size. The builders have patterns for several sizes, and are ready to make others as called for. Each size will break any stone, one end of which can be entered into the opening between the jaws at the top. The size of the machine is designated by the size of this opening ; thus, if the width of the jaws be ten inches, and the distance between them at the top five inches, we call the size ten by five.

The product of these machines per hour, in cubic yards of fragments, will vary considerably with the character of the stone broken. Stone that is *granular* in its fracture, like granite and most kinds of sandstone, will pass through more rapidly than that which is more compact in its structure. The kind of stone being the same, the product per hour will be in proportion to the width of the jaws, the distance between them at the *bottom* and the speed. The proper speed is about one hundred and eighty revolutions per minute ; and to make good road metal from hard compact stone, the jaws should be set from $1\frac{1}{4}$ to $1\frac{1}{2}$ inches apart at the bottom. For softer and for granular stones, they may be set wider.

The following table shows the several sizes of machines for which patterns are on hand,—the product per hour of each size, of fine road metal from the hardest materials when run with a speed of one hundred and eighty,—the power required to perform this duty, the whole weight of each size, in round numbers, and the weight of the heaviest piece when separated for transportation :—

SIZE.	Product per hour.	Power required.	Total Weight.	Weight of Frame and parts attached.	Cost, Jan'y, 1870.
10×5, . .	4 cubic y'ds.	6 horse,	6,600 lbs.	3,200 lbs.	$800 00
10×7, . .	4 "	6 "	7,600 "	4,100 "	900 00
15×5, . .	6 "	9 "	9,100 "	4,700 "	1,050 00
15×7, . .	6 "	9 "	10,200 "	5,600 "	1,200 00
15×9, . .	6 "	9 "	11,600 "	6,800 "	1,250 00

Since steam-engines, as usually rated by their makers, do not work economically to more than one-half their given horse power, an engine of double the power given in the table had best be bought.

The whole length of the machines to the backside of the wheels, is from eight to eight and one-half feet; height to top of wheels, five feet; width, from four to five feet.

The machine may be driven by any power less than that given in the table, yielding a product per hour smaller in the same proportion.

The proper sizes for making macadam, are the fifteen by seven and fifteen by nine for towns and parks. Public institutions and others requiring less quantities to be got out, can use the ten by seven size.

When broken by hand and for country roads, the stones should be broken on the storage places already mentioned, which are to be established along the side of the road every 200–250 feet. The laborer is not to pound the stones on a heap of such, but to use one large stone as a sort of anvil to break the others on. He is to use a light hammer, except for pieces containing more than four or five cubic feet, and may use a ring with a handle attached to hold the stone he desires to break.

In order that the road shall get an even surface the macadam must all be of one size, and the proper size for the macadam depends on the degree of hardness of the rock. If too small it turns to dust, if too large the top will not pack even. The size is regulated by the use of a ring as a gauge,—every stone being obliged to be capable of falling through this ring in any direction it may be dropped. Hard stones should be one to one and a quarter, softer ones one and a half and the softest two inches in diameter. Larger sizes give less perfect roads. In loading and otherwise handling macadam, a many and close-pronged pitchfork should be used instead of a shovel, so as not to mix in any earth or sand and to sift out the stone dust and chips.

The macadam being properly prepared and loaded up, it is spread over the foundation in two or three successive layers. Each layer *should* be rolled, but the top and last one *must* be rolled to make a good road. Nor will rolling alone do the work. Two other helps are needed: the use of a binding material, to act as a cement between the broken stone, and sprinkling. It is difficult to prescribe in words just what to use as binding material and just how much to sprinkle and roll; common sense will in most cases be a safe enough guide. In the macadamized streets of Paris the rule is to roll till a single piece of macadam placed under the roller, will be crushed, without being pressed into the road surface. Gravel somewhat mixed with clay by nature, but not too much, is probably best as a binding material. Clean coarse sand is very good. Other substances will do, where it would cost too much to procure either of the above.

The subject of rollers is one demanding some attention. In general, people are apt to over-estimate the value of a roller with respect to its weight. It will be evident on reflection that a roller should be as heavy per inch in length of roller, as a loaded wagon wheel is per inch of tire; or in other words, if we have a wagon with tires two and one-half inches wide and on each wheel a load of say one ton, the roller should weigh two-fifths ton for every inch in length, or a roller three feet long should weigh about fourteen and one-half tons, or else a wagon as above described would exercise more pressure on the road-bed per square inch than the roller and consequently would cut into the rolled surface and produce ruts. Road-rollers are of two principal kinds: those pulled by horses and those propelled by steam. The latter are for many reasons the best. In the first place they can be made as heavy as desired, without proportionally increasing the cost of propelling them, and being self-propelling, the only track they make is that of the roller, whereas with horse rollers, the hoof-marks of the horses are a great objection. Then again in the amount of work they will do at a certain cost, they excel horse rollers. They may be briefly described as a sort of locomotive mounted on three or four very broad and heavy wheels, these latter being the road-rollers.

There are several varieties in use in France and England and two at least of the English kind have been imported into this country, one for the New York Central Park, the other for the Arsenal Grounds in Philadelphia. The cost of the Central Park steam road-roller made by Aveling & Porter of Rochester, Kent, England, was about $5,000, set up in New York, and the amount of work it will do in one day at a running expense of $10, has been given as equal to that of a seven-ton, eight-horse road-roller in two days at $20 per day, or in other words, it will do the same work at one-quarter the running cost and in one-half the time, of a first-class horse road-roller.

The best horse road-roller of which the writer has any cognizance is the one shown by the annexed drawings in plan, elevation and in perspective. (See pp. 30, 32, 34.) It originates in Chemnitz, Germany, but can of course be easily made by any machine-shop or foundry. The hollow roller is made of cast-iron and is so arranged that it may be filled with water when it is to be used in heavy rolling ; when not in use and about to be moved from place to place, the water is allowed to run out, thus materially lessening the load. A circular cast-iron frame A, surrounds the roller, and carries the axle bearings of the same. The outside of this frame is turned to form a groove in which a strong wrought iron ring is fitted in such a manner that it will turn easily around the former. This wrought iron ring consists of two semi-circular parts, at whose junction the pole is attached on one side, and on the other an extension bar, carrying the balance weight *c*, which may be shifted by means of the set clamp *d*, or turned up by means of the hinge *b*. Pins going through the holes at *e*, fasten this ring or allow it to be turned for the purpose of pulling the roller in the contrary direction, when desired. The brake is shown at *f*, *f*, and consists of four wooden brake-blocks, attached by iron shoes to a bar behind them and having rubber packing between the shoes. The screws shown and the handles *h*, are used to operate these brakes.

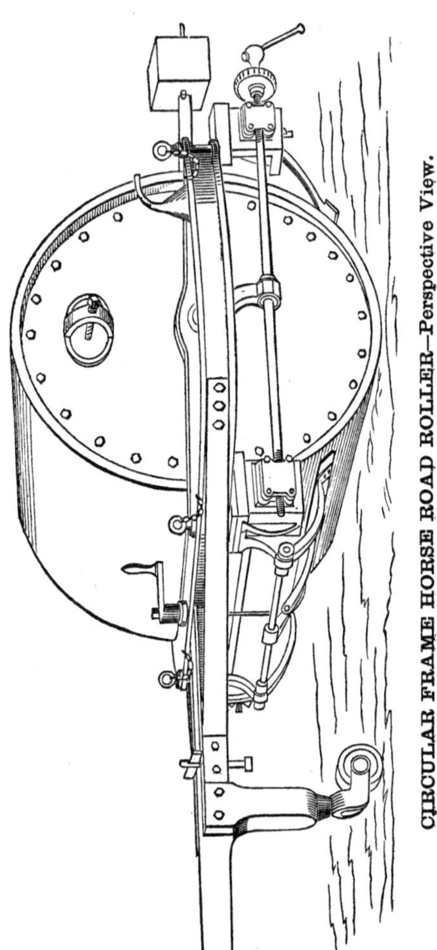

CIRCULAR FRAME HORSE ROAD ROLLER—Perspective View.

The cranks *m*, working the screws *n*, operate the scrapers *l*, which are used to keep the roller clean in muddy weather. The frame A, is made heavier at *o*, so as to have increased weight there to balance the whole frame-work in turning around. The support *p*, and the guide wheel *k*, might be dispensed with. A great saving in time and in movements hurtful to the road is effected by making the frame circular as described, this allowing the roller to be turned with the greatest ease. The dimensions are figured on the drawing. A roller of this kind four and one-half feet in diameter and three and one-half feet long, and weighing some four tons when empty, would cost perhaps $560–$600; one 5 feet × 3′ 8″, weighing about five and one-quarter tons (empty,) some $700–$750. Leaving off the break would diminish the cost about $50.

Before leaving the subject of macadam top roads, it ought to be mentioned, that a bed of rubble stone 10″ or 12″ deep, merely spread uniformly over the road-bed as a foundation, is better than nothing at all, but can never make the same quality of road as the rough paving described above.

The following data are to be used in estimating the cost of the kind of road just described. Rough foundation paving, pieces 5″–6″ long, filling up crevices and ramming the whole with hand rammers, costs after the material has been brought to the spot, one day's work of a common laborer for every four square yards, this assuming that the paver gets one and two-thirds common laborer's wages. Same kind of paving if set in sand will cost one day's work of a common laborer for every two and one-quarter square yards.

To make macadam by hand costs, for sizes from 1¼–1½″ of very hard rock, one day's work for every 0.6–0.44 cubic yards, for less hard rock, one day's wages will make 0.7–0.6 cubic yards, and of soft rocks 1.76–1.17 cubic yards.

To spread 14–12 cubic yards of macadam is also about a day's work.

Gravel Top.—Instead of the macadam top described in the preceding articles, screened gravel may be used. These roads are the favorite ones in Central Park, New York, and are probably the best road there is for pleasure drives. It is a matter of some doubt yet whether they do as well for heavy trucking as they do for light vehicles. The foundation for these gravel roads

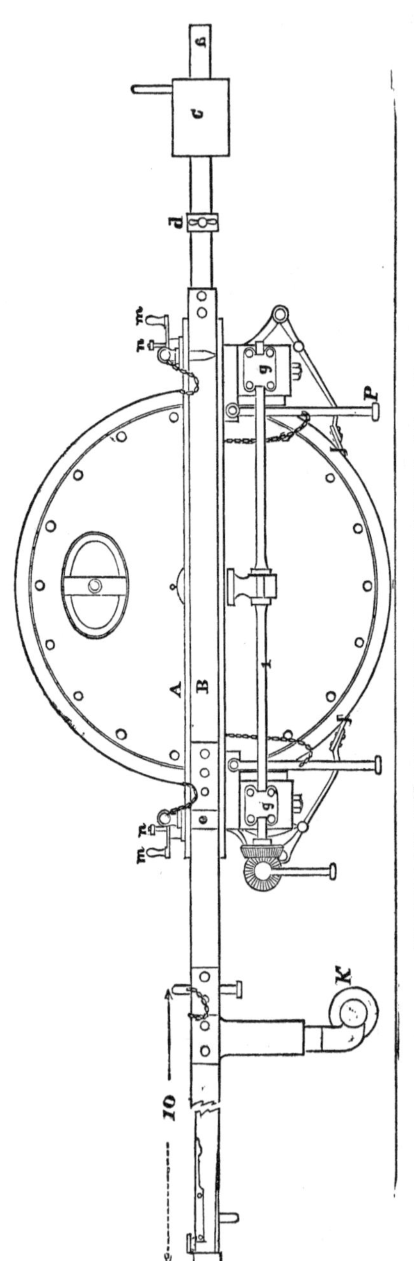

CIRCULAR FRAME HORSE ROAD-ROLLER.—Elevation.

should be the same as the rough paving for the macadam road; some pieces were built in the Central Park, having a rubble stone foundation but they are not recommended by their builders. The gravel to be used for the top must be selected with some care; it should be of a hard kind of stone, clean, that is, free from clay, &c., of the right color, &c. It is put on in two layers, each rolled, and the top one made compact and firm, by spreading and mixing in some good binding material, sprinkling and rolling. There need be no fear of making a poor road by using the smoothest, most water-worn pebbles, free from all sand, &c., in making a road-top. The upper portions of the river Rhine are remarkable for the clean, smooth pebbles that form its bed to a very great depth. These pebbles are dredged up and used in road-building, making an excellent road-covering at a small expense. There are many miles of such roads in Baden and in the Bavarian Rhine Provinces.

Keeping Roads in Repair.—This subject properly finds its place here, being a matter of skill and a thing of debate only in the case of what we have called foundation roads; pavements and trackway roads, to be considered after this, need no special directions as regards their repair or maintenance.

After a road has been properly rolled, and the surface made compact and smooth, it should always be maintained in that condition, no matter how great is the amount of travel on it. "A stitch in time saves nine," here as well as elsewhere. The tendency is to produce ruts; these gather water; this soaks up the road-bed and spoils the whole. The problem can be put in this way: To have a good road, it is necessary that there be no dust or mud on the same, and that there be no ruts; therefore remove the dust and mud as fast as they are formed, and fill up the ruts as fast as they are made. The whole matter is here in a nut-shell. It may be thought, at the first view, that this is too expensive a system. Its principal beauty lies, however, in the fact that it costs less per mile of road kept one year than the pernicious system of annual or semi-annual repairs, as will be shown and proved. The above two rules—sweep off the mud and dust as fast as they are formed, and fill up the ruts and bad places with new material as fast as they appear—are all that is necessary to be carried out in order that there be *continually* a good road. Without continual repairs, there can be no

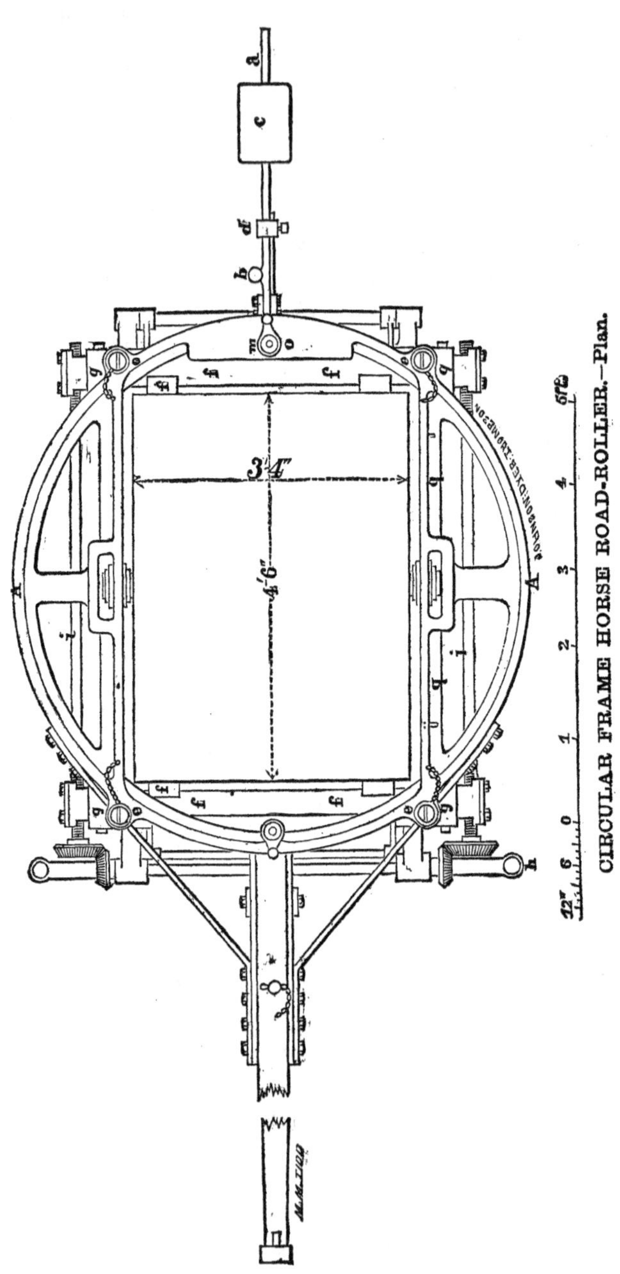

CIRCULAR FRAME HORSE ROAD-ROLLER.—Plan.

such thing as a constantly good road—a proposition that cannot too often be repeated. By repairing a road annually, or twice a year, it matters not which, the result is, strictly speaking, a *good* road at NO time during the whole year. The road is wretched just after repairs; it becomes *passable* after a while, and deteriorates from that day forward, until it is again made wretched; and so on, *ad infinitum*, according to the present only too commonly followed system. By the other method is offered us a road as smooth as a floor, year in, year out, and, let it not be forgotten, *at a less expense*.

A French engineer, named Trèsaguet, was the first, in 1775, to call attention to this proper method of making road repairs. His system—the above described one—was adopted in Baden in the year 1845, and has been long in universal use in all the active European countries. The two tables below give, the first, the actual average quantity of road macadam used per mile of road in Baden to make the repairs in one year, and show the decrease after 1845. The second gives, in the first column, the cost of materials and labor required to repair one league for one year according to the old way,—this column being calculated for the years following 1845 from the cost of the preceding years, and allowing for the increased value of labor and materials,—while in the second column we have the actual cost, as it was with the system followed at the time:—

TABLE I.

YEAR.	Cubic yards used per mile in one year to repair roads.
1832,	218.6
1839,	198.7
1851,	127.2
1855,	91.4
1856,	89.4
1860,	93.4

TABLE II.

YEAR.	Cost of Repairs of one league of Road.	
	By old way of so doing, in florins.	By system of continual repairs, in florins.
1835, ...	1,002	1,002
1840, ...	1,086	1,086
1845, ...	1,170	975$\frac{36}{60}$
1850, ...	1,254	965$\frac{40}{60}$
1855, ...	1,339	835$\frac{5}{60}$
1860, ...	1,423	978$\frac{54}{60}$

These figures are taken as given by the chief engineer of the Baden Public Works, Mr. Keller. He quaintly adds: "These tables give clear evidence in favor of the reduced *cost* by the adopted system. That roads are *better* now than they formerly were, everybody knows." Another German engineer expresses himself to the same effect in a little different way. "It costs no more," says he, "to keep the roads in repairs now (1864,) than it did twenty years ago, when this method (of continual repairs) was not in use, although labor is now three times and materials are twice as dear as they then were." There seems to be no doubt of the superiority of the continual repair system in every respect, producing very much better roads, and at the same time costing less. It need only be tried with us to be thenceforth adopted.

HOW TO REPAIR ROADS ON THE CONTINUOUS SYSTEM.

We suppose the material for the road-covering to lie in regular measured heaps, all ready to be used, at the storage places, once or twice above mentioned, as being 200–250 feet apart alongside of the road, but not encroaching upon it. Then, for every two or three miles of road, a so-called road-keeper is employed to do the necessary work and repairs. An enumeration of his duties will comprise at the same time an essay on the art of road repairing.

1. The road-keeper is to remove the dust formed in dry weather by sweeping with a brush broom. This is done to greatest advantage just after a slight shower. In muddy weather, it is essential that the mud be removed by means of brooms or hoes. A little mud on the surface causes ruts, and much mud softens up the whole road-surface. The mud is to be raked up in heaps alongside of the road, there left to dry and then carted off. To hinder as much as possible the formation of any mud, the surface drainage must remain unimpaired; should it be out of order, the water standing on the road is to be swept off. To diminish the wear of the road in dry times, the road should be sprinkled.*

2. Inasmuch as the covering gradually wears off, notwithstanding all precautions, it must be renewed, and should be so renewed gradually, in the same measure as it wears off. The best time to put on new road metalling is during continuous wet weather.

3. In filling up holes, the bottom of the same is to be swept clean of mud, then filled up level with the remainder of the road, not in a heap so high above it as to obstruct travel.

Every care should be taken to have the new material join as speedily as possible with the old portion of the road, and it should be so well laid that it will give the least possible hindrance to vehicles, which will then not avoid the patched places.

4. When many ruts occur in a short distance, the deepest only are to be filled at first. After the patching in these has become solid, then the rest are to be attended to. Long ruts or wheel tracks are not to be filled up the whole length at once, but only short pieces at a time. If this precaution is neglected, vehicles avoid such places, and new ruts are formed elsewhere.

5. Inasmuch as more material is worn off in a dry season than can be put on, there are then, when wet weather comes, large places to be repaired. These must be mended by degrees, never filling up a piece larger than 8–10 feet × 4–7 feet, at a time and not having these pieces too near together; when these have

* Bowles, in his book, "Our New West," mentions the case of the stage road from Sacramento to Virginia City, *via* Placerville, one hundred and fifty miles long, and having an annual traffic of seven or eight thousand heavy teams, and whose proprietors found that the simplest and cheapest way of keeping it in repair during dry weather was to sprinkle the whole of it,—one hundred and fifty miles of mountain road.

become solid, then some more may be fitted in and so on till the whole is done.

Should it however become absolutely necessary to repair a piece of road in dry weather, the place where the new macadam is to be deposited must be loosened up with a pick, then the new material put on and a solid top formed by the judicious use of stone dust or other binding material and sprinkling with water and pounding down with the shovel, or by what may be called "puddling" until the whole be solid. Should a frost or very dry weather occur immediately after macadam has been put on the road in wet weather so that the same will not join on the rest of the road surface, the whole must be removed, cleaned and returned to the storage heaps for future use. A layer of macadam over the whole road should never be put on, without treating it immediately afterwards in the manner described above for building new roads, that is, mixing in binding material with the top course and rolling it in wet weather or after sprinkling.

The road-keeper is naturally also the person to see to the proper delivery on the part of the contractors, if such there be, of the road material in the prescribed places and to attend to the measuring of the same.

In short and to sum up, it is his business to keep the road in good order, and with proper men and surveillance the desired result is achieved easily and at a less cost, than by any other system. The quantity of macadam required to keep a certain length of road in repair varies very much; it depends as we have seen on the care with which the repairs are made, naturally also on the kind of stone used and on the amount of travel over the road. For a width of road=twenty feet, the average quantities required per year to keep a length of ten feet in repair, on the system of continuous repairs, has been given as follows:—

	Cubic ft.		Cubic yds.
1. Good material and heavy travel, ...	. 15–20	=	.55–.74
2. Good material and medium amount of travel, .	. 10–15	=	.37–.55
3. Good material and light travel, ...	. 5–10	=	.18–.37
4. Medium material and heavy travel, ...	. 20–25	=	.74–.92
5. Medium material and medium amount of travel, . .	. 15–20	=	.55–.74
6. Medium material and light travel, ...	. 10–15	=	.37–.55
7. Third rate material and heavy travel, ...	. 25–30	=	.92–1.01
8. Third rate material and medium amount of travel,	. 20–25	=	.74–.92
9. Third rate material and light travel, ...	. 15–20	=	.55–.74

These are the quantities as given by one authority, but from a comparison with the amounts actually used during a period of ten years on thirty-nine roads, having very various amounts of travel upon them and being repaired with all kinds of road metal, it would seem that the above figures are very ample.

Repairs of Macadamized and much frequented Streets in Cities.

In this case, where the amount of travel in one day is often greater than that of a month or more on the town road, the system of continuous repairs ceases to be the best available, on account of the incessant throng of vehicles not giving any repaired place a chance to become solid before it is again ploughed up and scattered. Thus in the city of Paris on the Boulevards, &c., the continuous system has been abandoned and the practice now is to let the street gradually wear down three to four inches, then close half of it (divided "fore and aft") to travel, loosen it all up with picks and put on a layer three or four inches [best not to put on more than that] spread a thin layer of sand over this, sprinkle and roll heavily. It often happens that the men put too much of the sand on; in that case, the road, after it is all done, is finally well watered and the roller again passed over it a number of times. This operation causes the superfluous binding material to come to the surface in the shape of thin mud and leaves the road covering as hard and smooth as mosaic, making a most excellent drive-way. It emits a sonorous, ringing sound on being driven over and remains clean and without mud throughout the heaviest rain-storms. The rolling of the streets in Paris, is done by a company owning a large number of steam rollers; in paying them for work done, the city was obliged to go back to first principles for a measure of such work, it being found impossible to estimate correctly by the square measure of surface rolled to such and such a degree of hardness. The measure adopted is that of weight multiplied into the distance it has been moved, or "feet pounds" as we should say. It has been found from many years experience that to roll one cubic meter of macadam requires 4–5 "Kilometer-tonnes" and this is true whether the layer of macadam be three and one-quarter or ten inches thick. Expressed in our measures this is 11,020–

13,775 feet tons @ 2,000 lbs.=2.09–2.61 mile tons per cubic yard of macadam.

Pavements and Trackways.

No essay on roads would be complete without some mention of these two species of road-surface, though the use of the former is confined principally to streets, and that of the latter is out of date.

Pavements are either of stone, wood, iron, various concretes, asphalt, and may be of still other substances.

Stone Pavements.—The modern sizes of paving stones may be seen from the following cases. The Boston size is $4\frac{1}{2}'' \times 3\frac{1}{2}'' \times 7''$ deep; New York Belgian, 6–$8'' \times 5$–$6'' \times 6$–$7''$ deep; new Broadway pavement, also called Guidet pavement, $3\frac{1}{2}$–$4\frac{1}{2}'' \times 10$–$14'' \times 7\frac{1}{2}$–$8\frac{1}{2}''$ deep. This last is laid with the long sides of the stones across the street; and, as far as the author's judgment goe , is the best size for stone pavement there is. The Boston size is too small, and allows of no bond between the separate paving stones. Further, the weakest part of each stone being its edge, it follows that the more edges there are in a given surface of pavement, the speedier will it wear out, each stone becoming rounded and slippery. It is only the excellent workmanship and great care displayed in setting these stones in Boston that prevents these facts from being at once apparent to all. When it is added that in setting pavements, the natural soil, except it be sand or fine gravel, is in all cases to be excavated 12–19 inches, and then filled up 5–12 inches, according to the solidity of the subsoil, with *clean*, coarse sand or fine, *clean* gravel, and the paving stone set in this and well rammed down with hand rammers, about as much is said on this topic as can be said without going into long details.

From four and one-half to six cubic feet of sand are required for every square yard of paving. In setting two different pavements, the same written rules may be exactly followed in either case, yet one be much better than the other, so much depends here upon good, careful, conscientious workmanship.

Wooden Pavements.—There are so many kinds of these, that it would be out of place to enumerate and describe them all here. Their advantages are, less wear on tires and horses, less noise and smooth traction; a disadvantage, is their slipperiness in the winter. There seems to be a sort of notion that wood pavements and coal tar must go hand in hand; but there certainly is no necessity for this. Coal tar is applied as a preservative to the wood; but it must be acknowledged that many better ones are known and indeed are used, to the utter exclusion of coal tar, in all cases where it is desired to preserve wood, except in this of wood pavements. No wood should be used in paving that has not been first subjected to some approved method of preservation, or impregnation, as it is frequently called. The best manner of setting the same is still a mooted point, which it would be presumptuous at present to decide.

Cast-iron pavements are out of favor on account of their great cost, and concrete pavements are a matter of experiment as yet.

Asphalt pavements are chiefly used in Paris. They are slippery in wet weather, and produce a very disagreeable, penetrating dust in dry weather. It is necessary to prepare a bed of macadam to lay them on, and they are not used in Paris except in streets where the gas-pipes are carried either in the sewers or under the sidewalks, as any leak of gas would destroy them. Their use is a matter of doubtful economy.

Trackways are, as has been mentioned, out of date. Where a common road does not suffice now-a-days, a railroad is built; but time was when trackways were of considerable importance. They consist, if of stone, of large, flat stones, say 12″ deep and 4–6 feet long by 14″–16″ wide, solidly bedded in two parallel rows, at such distance apart as to make of each row a track for the wheels. The space between is paved. They are of course very expensive, but cost little to repair, and enable a horse to pull a very great load. As has been mentioned, Telford made use of such a stone trackway, to avoid cutting down a hill, on his Holyhead road. There were two hills, each a mile in length, with an inclination of 5 in a hundred. It would have cost $100,000 to reduce this grade to $4\frac{1}{6}$ in a hundred, but nearly the same advantage, in diminishing the tractive force required, was obtained by keeping the 5 in a hundred grade, with moderate cuttings and embankments, and making stone trackways, at a total expense of less than half the former amount.

"Plank roads," once so much in vogue in the United States, may not improperly be classed among roads with trackways, and, with them, also among the things that were. From their perishable nature, they can never advantageously do more than help the development of a new country, and in this, as well as other States, are yearly becoming more and more impracticable on account of the constantly increasing price of lumber.

On the Resistance to Motion or the Force required to move Vehicles on different Kinds of Roads.

Before, as well as since the introduction of railways, engineers in England, Germany and France made many experiments on the force necessary to pull different vehicles, at various speeds over various surfaces. To enumerate the details of all these experiments would be perhaps useless; a few general results only are here given.

Experiments, as above indicated, were made by Edgeworth, Count Rumford, Bevan, Macneill, Minard, Navier, Perdonnet, Poncelet, Flachat, Morin, Kossak, Umpfenbach, Gerstner, and no doubt others, a list of authorities that proves the subject to have been well nigh exhausted. The experiments of Morin, made in 1838–41, appear to have been made with a degree of care and accuracy, leaving nothing more to be desired, and the following table is an extract from his results,* and gives that fraction of the weight of the vehicle and load, which is required to move them on a level road:—

* A full account of Morin's experiments on the resistance to motion of vehicles, on the wear caused by different vehicles on roads and on the loads different vehicles should carry so as to produce the same wear, may be found in Morin, *Expèrience sur le tirage des Voitures, Paris*, 1842.

CHARACTER OF THE ROAD.	CHARACTER OF THE VEHICLE.					
	2-wheeled carts.	Trucks, 4-wh., three and four horse.	4 horse stage-coaches, on springs.		2-horse carriages, body on springs.	
Firm soil, covered with gravel 4″–6″ deep, ...	$\frac{1}{12}$	$\frac{1}{9}$	$\frac{1}{8}$		$\frac{1}{8}$	
Firm embankment, covered with gravel $1\frac{1}{4}''$–$1\frac{1}{2}''$ deep, ...	$\frac{1}{16}$	$\frac{1}{11}$	$\frac{1}{10}$		$\frac{1}{10}$	
Earth embankment, in very good condition, ...	$\frac{1}{41}$	$\frac{1}{29}$	$\frac{1}{26}$		$\frac{1}{26}$	
Bridge flooring of thick oak plank, .	$\frac{1}{70}$	$\frac{1}{46}$	$\frac{1}{41}$		$\frac{1}{42}$	
BROKEN STONE ROAD.			Walk.	Trot.	Walk.	Trot.
In very good condition, very dry, compact and even, ...	$\frac{1}{75}$	$\frac{1}{54}$	$\frac{1}{48}$	$\frac{1}{41}$	$\frac{1}{49}$	$\frac{1}{42}$
A little moist or a little dusty, . .	$\frac{1}{53}$	$\frac{1}{38}$	$\frac{1}{34}$	$\frac{1}{27}$	$\frac{1}{34}$	$\frac{1}{27}$
Firm, but with ruts and mud, . .	$\frac{1}{33}$	$\frac{1}{24}$	$\frac{1}{21}$	$\frac{1}{18}$	$\frac{1}{22}$	$\frac{1}{19}$
Very bad, ruts 4″–$4\frac{1}{2}''$ deep, thick mud,	$\frac{1}{19}$	$\frac{1}{14}$	$\frac{1}{12}$	$\frac{1}{10}$	$\frac{1}{12}$	$\frac{1}{10}$
Good pavement, Dry, ...	$\frac{1}{90}$	$\frac{1}{65}$	$\frac{1}{57}$	$\frac{1}{38}$	$\frac{1}{59}$	$\frac{1}{39}$
Good pavement, Covered with mud, .	$\frac{1}{69}$	$\frac{1}{50}$	$\frac{1}{44}$	$\frac{1}{33}$	$\frac{1}{45}$	$\frac{1}{34}$

To take an example, suppose we have a truck weighing with its load 9,000 lbs. How many pounds traction will be required to move the same?

Ans.—On firm soil, gravel 4″–6″ deep, that is, a newly repaired road, as we often find it, ($\frac{1}{9}$ by table,) 1,000 lbs.; on best kind of embankment, ($\frac{1}{29}$ by table,) 310.3 lbs.; on broken stone road in good condition, ($\frac{1}{54}$ by table,) 166.6 lbs.; on broken stone road, deep ruts and mud, ($\frac{1}{14}$ by table,) 643. lbs.; on a good pavement, ($\frac{1}{65}$ by table,) 138.5 lbs. Or, since the tractive force of a medium horse when working all day is said to be about 125 lbs., we need in the first case, 8 horses; in the second case, $2\frac{1}{2}$ horses; in the third case, about $1\frac{1}{4}$ horses; in the fourth case, about 5 horses; and in the fifth case, only one good horse to move the same entire load all day.

These facts expressed in the preceding page or two in striking, yet perhaps dry figures, can be nearly as well given in popular language.

Says a correspondent (Dr. Holland,) of the "Springfield Republican," writing from England, after describing the kind of horses in use there:—

"Now with all these horses the rule follows that every pound of muscle does just as much work on the road as two pounds do in America. The cab and omnibus horse does twice as much as the same horse does in America. The draft horse does as much at the dray as two ordinary dray horses in America, and the little horses, which are driven mainly in butchers' carts and grocers' carts, will tire a cab horse to follow them with no load at all.

"In connection with these statements it should be recorded that the speed of all vehicles in the streets of London, whether the localities be crowded or not, is at least a third faster than it is in corresponding streets in American cities. The ordinary speed of vehicles in London, in which passengers or light loads are transported, is one which is considered not entirely safe in Main Street, Springfield, Mass., and one which, in some streets of Boston or New York, would be at once checked by the police. A man who sits in a 'Hansom' finds himself driven at an unprecedented pace through crowded thoroughfares, and Yankee though he may be, he will often wonder whether he is going to bring up at last without a broken neck.

"I mention this matter of speed, particularly, because it shows that even more work is done by one horse in London, than by two in New York. He not only draws as large a load, but he travels with greater rapidity. The streets of London present such a spectacle of headlong activity as no American city can show, in consequence of the rapid passage of all sorts of vehicles through the streets. I might add to this statement, touching the superior speed of the London horses, a word about the greater weight of the carriages which they are obliged to draw behind them. All carriages are built more heavily in Great Britain than in America. They are built to last, and many of them seem to me to be superfluously heavy.

"The point which I wish to impress upon my American reader is simply this: that the English horse, employed in the streets of a city, or on the roads of the country, does twice as much work as the American horse similarly employed in America. This is the patent, undeniable fact. No man can fail to see it who has his eyes about him. How does he do it? Why does he do it? These are most important questions to an American. Is the English horse better than the American? Not at all. Is he overworked? I have seen no evidence that he is. I have seen but one lame horse in London. The simple explanation is that the Englishman has invested in perfect and permanent roads what the American expends in perishable horses that require to be fed. We are using to-day, in the little town of Springfield, just twice as many horses as would be necessary to do its business if the roads all over the town were as good as Main Street is from Ferry to Central. We are supporting hundreds of horses to drag loads through holes that ought to be filled, over sand that should be hardened, through mud that ought not to be permitted to exist. We have the misery of bad roads, and are actually or practically called upon to pay a premium for them. It would be demonstrably cheaper to have good roads than poor ones. It is so here. A road well built is easily kept in repair. A mile of good macadamized road is more easily supported than a poor horse."

Other results of Morin's experiments are as follows:—

1. The force required to draw a vehicle, is directly proportional to the load and inversely so to the diameter of the wheels; in other more common words, the tractive force increases in the same ratio that the load increases, and the diameters of the wheels decrease.

2. On a paved or well built macadam road, the tractive force is independent of the width of the tires, provided the same is more than three or four inches. On compressible roads, such as new gravel, on a meadow, &c., the tractive force diminishes with an increase in the width of the tires.

3. Other circumstances being equal, the tractive force is the same for vehicles with and without springs as long as the horses are not moving faster than a walk.

4. On paved and well macadamized roads the tractive force increases with the velocity, according to the law, that beyond a velocity of 2¼ miles per hour (23.31 feet per second) the increase of the tractive force is in direct proportion to the in-

crease in velocity; this increment is however less, the softer the track or road and according as the vehicle is best provided with springs.

5. On soft earth embankments, or on sand or sods, or on streets newly covered with gravel, the tractive force is independent of the velocity.

6. On a well made pavement of regular shaped stone, the tractive force, horses on a walk, is about three-fourths of that on a good macadam road, but with horses on a trot, the two are about equal.

7. The wear on the road is greater the smaller the diameter of the wheels and greater in the case of vehicles without, than for those with springs. Most road-rollers, as now in use, have too small a diameter besides being too light and consequently do not properly compress the road surface.

8. The tractive force, as well as the wear on the road, is greater in the case of vehicles that have their wheels placed at an angle with the vertical by reason of the ends of the axle-trees being bent down, than for those that have their wheels set plumb and the centre line of the axle-trees level.

PART II.

ON THE "BEST METHODS OF SUPERINTENDING THE CONSTRUCTION AND REPAIR OF PUBLIC ROADS IN THIS COMMONWEALTH."

In looking for a solution of this question the people of the Commonwealth may turn as they choose, either to the West or to the East, to see a guiding star; to the city of Chicago, or to the city of London, both under a republican form of government, alike or similar to that we live under. It lies in the establishment of a Board of Public Works, composed of a number of able men, well paid for their services, gradually changing in their membership in the Board who shall have this and only this as their occupation and who can therefore be held responsible for their acts. This is the system that has been adopted both in London and in Chicago and with remarkable success and resultant benefits. There are many other systems in use in foreign countries all of which however seem to be inapplicable here, placed as we are, under so different forms of government; hence, though well acquainted with the systems adopted in France and in Germany, the writer has not described them here.

The history of "the Metropolitan Board of Public Works of the City of London" is about as follows:—

What is known as the city of London consists in reality of a great number of what we should call towns, there called parishes, and of which the "*city of London*" is only one single member. Each one of these parishes had, and still has in most respects, its own local government and in consequence, took care of its drainage, its streets, &c., &c., as seemed best and as it liked, some better, some worse and some not at all. This state of things in the matter of drains and sewers finally led to a most deplorable condition of affairs; there was not nor could there under these conditions be such a thing as a *system* of sewers and consequently a proper and adequate drainage; the death-rate increased to an alarming extent and matters came to be universally regarded as past all eudurance. What could be the remedy? No well grounded complaint could be made against the majority of the men composing the various local governments, since they were good and honest citizens and hence no change in the separate governments could ever bring relief. The fault lay not in the men, but in the system of ruling they were called upon to fulfill, that is, in the incompetent and faulty tread-mill of government they were annually called upon to keep in its usual operation. It was then seen that by having an elected power to supervise and regulate the sewage affairs of the whole metropolis, a complete *system* of drainage could be carried out, and thus only. Such a regulating power is exercised by the metropolitan Board of Public Works, chartered by Act of Parliament and composed of members elected from all parts of London. It is perhaps in place here to explain what is meant by a *system* of sewers as the same definition will hold good in other matters; as for a *system* of roads, of drainage and irrigation of lands, &c. Perhaps the best illustration would be to refer one to the veins and arteries in the human body, or to the body of a tree, from its trunk through the branches growing smaller and smaller down to the smallest twig that may be on it. It will be at once seen how different any arrangement, in which may be detected the wisdom to contrive, the strength to uphold and the beauty to adorn, like this, is from a miserable patchwork such as cannot but arise where the separate parts of one whole are each left to guide themselves without any unity of action or design, as to their final resultant. The London Board of Public Works had some extraordinary powers conferred upon it, such as the right to levy assessments on real estate benefited by their improvements, and others. Originally constituted merely to plan and execute a system of sewerage for the metropolis, this Board of Public Works soon showed itself so useful and beneficial in its actions that other matters were placed in its charge, such as the laying out of new streets, the building of the Thames embankment,—a work of exceeding great magnitude and importance,—and there seems to be no doubt that in all public works London will find it advantageous to employ its Metropolitan Board of Public Works.

In the city of Chicago there has been a Board of Public Works almost from the very start. It arose there from the union of the water supply and the sewerage commissioners, and has existed since May, 1861. No less than in London, it has proved to be of great benefit to the community; and it would have been impossible, under any other system, to have executed in so satisfactory a manner the many and useful public works for which Chicago is famed. At the risk of introducing in this place some very dry reading, a general synopsis of those parts of the city charter which relate to the Chicago Board of Public Works is here given. The whole may be found in a copy of "Laws and Ordinances, Chicago, 1866."

SECT. 1. Establishes a body known as the *Chicago Board of Public Works*, to consist of (3) three members, chosen by the people, one from each division of city.

The first *three* chosen for one, two and three years; after that, one each year for three years.

SECT. 2. Each member of board shall receive annual salary of three thousand dollars (by Act of February, 1866); give bonds for faithful discharge of duties; pay over all moneys, papers, &c., at expiration of his term, or when ordered by city council.

SECT. 3. Board to elect president and treasurer, and make by-laws.

SECT. 4. Majority constitutes quorum; records to be kept of proceedings; copies of all plans, estimates, &c., to be kept; report (annual,) to be rendered on or before each year, or when required by city council. Each member authorized to administer legal oaths.

SECT. 5. Board shall have special *charge* and *superintendence*, subject to the laws and ordinances of city council, of all streets, lanes, alleys, &c., in the city of Chicago, and of all walks and crossings in the same, and of all bridges, docks, wharves, public places, landings, grounds and parks in said city, and of all halls, engine-houses and other public buildings in the city belonging to city, except school-houses, and of the *erection* of *all* public buildings; of lamps and lights in streets, &c., and in public buildings, and repairs of same; of the harbor works and improvements; of the city sewers and drains and of the water works; of the fire-alarm telegraph, and all public works and improvements hereafter to be commenced by the city, as well as such other duties as may be prescribed by the city council by ordinance.

SECT. 6. All applications or propositions for improvements,

7

or new works of kind specified in section five, shall hereafter be first made to Board of Public Works, or if made first to city council, shall be by them referred to board. Upon receiving application, board shall investigate the same, and if they find such work necessary and proper, shall thus report to city council, with an estimate of the expense thereof. If they do not approve of such application, they shall report the reasons for their disapproval, and the city council may then, in either case, reject said application or order the doing of work or making of public improvement, after having first obtained plans and estimates thereof. The board may also in like manner recommend, whenever they think proper, any improvement of the nature above specified, though no application has been made therefor.

SECT. 7. Shall be duty of board to procure for city full plans and estimates of contemplated improvements, when so ordered by council.

SECT. 8. Whenever any public improvement shall be ordered by city council, and money appropriated, board shall advertise for proposals for doing work; plans and specifications of same first placed on file in office of board, which plans and specifications shall be open to public inspection; advertisement to state work to be done, and to be published ten days at least. The bids shall *be sealed bids*, directed to board, and accompanied by bond to city, signed by bidder and two responsible sureties, in sum of two hundred dollars, conditioned he shall do work if awarded to him; in case of his default to do so, &c. Bids to be opened at time and place mentioned in advertisement.

SECT. 9. All contracts shall be awarded to lowest reliable bidder, and who sufficiently guarantees to do work under superintendence and to satisfaction of board: *provided*, that the contract price does not exceed the estimate, or such other sum as shall be satisfactory to board. Copies of contracts to be filed with city comptroller.

SECT. 10. Board reserves right, in contracts, to decide questions as to proper performance of work and meaning of contracts; in case of improper construction may suspend work and relet same, or order entire reconstruction; or may relet to other contractors and settle for work done, &c.

In cases where contractor properly does work, board may, in their discretion, as work progresses, grant to said contractor estimate of amount already earned, reserving fifteen per cent. therefrom, which shall entitle holder to receive amount, all other conditions being satisfied.

SECT. 11. In case prosecution of any public work be suspended, or bid be deemed excessive, or bidders be not responsible, board may, with written approval of treasurer, where urgency of case and interests of city require it, employ workmen to perform or complete any improvement ordered by council: *provided*, that the cost and expense shall in no case exceed the amount appropriated for the same.

SECT. 12. All supplies of materials, &c., when costing over five hundred dollars, to be purchased by contract, subject to same conditions as letting out work.

SECT. 13. Whenever board think necessary for interests of city, to protect same from damage or loss, shall report thus to aldermen, and reasons for same, asking power to give contracts without notice required above, and aldermen may grant request: *provided*, three-fourths vote for it.

SECT. 14. Whenever board is of opinion work may be better done without contract, shall so report to council, and same may authorize board to procure machinery, materials, &c., hire workmen, &c.: *provided*, a three-fourths vote be in favor of granting authority.

SECT. 15. All contracts and bonds by board to be in name of city.

SECT. 16. No member to be interested in any contract; all contracts made with any member interested, city may declare void; any member so interested shall forfeit his office and be removed therefrom; the duty of every member of board and of every officer of city to report delinquency, if discovered.

SECT. 17. All existing contracts executed by city, by water or sewarage department, &c., to be carried out by board.

SECT. 18. Board shall nominate each year the various officers, now provided for by ordinance, which serve in the departments under their special charge, the city engineer, superintendent sewers, streets, &c. Shall be empowered to employ from time to time such other superintendents, clerks, &c., as they may deem necessary, subject to ordinance as regards pay, &c.

SECT. 19. Board to have charge and superintendence of works made for city, and paid for by private individuals or by State. Plans for same to be approved by board.

SECT. 20. Board shall, on or before every year, submit to auditor, by him to be presented to council with annual estimate, statement of the repairs and improvements necessary to be undertaken for current year, and of the sums required by board therefor; report to be in detail; report, having been revised by council, sums required shall be provided for in annual tax-levy. All moneys to be paid to any person out of moneys so raised, shall be certified by president of board to auditor, who shall draw warrant on treasurer therefor, stating to whom payable and to what fund chargeable; such warrant to be countersigned by president of board.

SECT. 21. Board to keep accounts, showing moneys received and spent, clearly and distinctly, and for what purpose. Accounts to be always open for inspection of auditor or any committee appointed by city council.

The object of introducing this synopsis here has been to give a complete picture of just what such a Board of Public Works is. It will be seen upon a little examination how entirely different a thing it is from the usual and only too customary "committee." Perhaps the greatest fault of a committee is its entire lack of what might be called body *and* soul. If corporations, as has been said, have no souls, a committee may be said to have neither body nor soul. It is alive to-day, wields great power, decides vital and important questions, and yet is nowhere to-morrow, and seemingly even its component atoms have vanished from the face of the earth. It is amusing and yet sad, when the action of some such committee has caused trouble to read some time after, that it all "is exceedingly discreditable to whoever is responsible for it." How much better to have a conservative, expert and reliable body, the members of which have *no other business* than to attend to their duties as such, who are well paid for it and consequently can at any time be held strictly responsible for their actions. With such a power, wisely governing and regulating the roads of this Commonwealth, it would be an easy matter to make thorough improvements in the legislation concerning roads and in the roads themselves.

These are two changes the need of which is generally felt at present and has found expression in various ways.

It may be well to quote one at least, notable for saying very much in little compass,—of these calls for improvement, in this connection, and adding some more as belonging to this subject in the form of an interesting appendix. Says Gov. Claflin in his Inaugural: "Few things are of greater importance to a community, or a surer test of civilization, than good roads. Those of our citizens who have visited Europe are unanimous in the opinion that our public roads are far inferior to those of other countries, where the means of easy and safe communication are better appreciated. The science of road-making is apparently not well understood; or, if it is, the present modes of superintending the construction and repair of roads are so defective that the public suffers to an extent of which few of us are aware. It may be found upon investigating the cause of our miserably poor and ill-constructed roads, that the laws relating to this subject need revision, so as to give more uniformity in their construction and the repair of our highways. It is evident, also, that the science of road-making should have a prominent place in the course of applied mathematics at the Massachusetts Agricultural College."

We stand then in this matter of roads at precisely the same point that the good people of London did ten or a dozen years ago in the matter of their drainage, and our remedy is the same. The fault lies in the machinery of government; originally built up to cater to the wants and needs of a newly settled country,—a colony breaking a path through the wilderness,—it has long since ceased to satisfy the demands of the present *State* in no matter so essentially as in that of its government and laws relating to common roads and highways. This is a subject requiring special knowledge, to be acquired only by long experience or the shorter method of imbibing the experience of others, which on analyzing it, is all that any *study* amounts to; formerly it was not so, and most any one sufficed to make improvements on Indian paths. We need then an *expert* government on this point.

There should be a distinction made between first, second and third class, or between, as they might be called, State, County and Town roads; the first two should not be left to be dealt with as it is the pleasure of each town. A chain cannot be perfect unless every link in it is so; no more can a road. The State must attend to the State and County roads and set a proper example at least to be followed by the towns in the case of their roads. We need then a higher power than that of the towns.

It has been previously shown how we need a power that can be held responsible and is somewhat permanent, and to put it all together, we need, to order and maintain our highways, a Massachusetts Board of Public Works. For some years it would have its hands full in improving the existing main roads and laying out some new ones, but in course of time, as in the older countries of Europe, its principal business would be the *maintenance* of the roads. It must be remembered that the Board of Public Works is merely the intelligent servant and adviser of the legislative and executive; whatever sums the legislature appropriates for certain objects, that is taken by the board and made to yield its most in the shape of work accomplished. Beyond this and keeping its accounts, it has nothing to do with money or taxation.

The small state of Baden, a part of Germany, has been heretofore mentioned as a model in road construction and the care of the same. From a brief history of the roads of that country and their present management, we may take some useful notes. The account is that of the Chief Engineer of the department of "Roads and Hydraulic Engineering," which has this matter in charge and is therefore reliable.

"In Baden the condition of the roads has been a subject of great care. Within the last forty-five years many millions have been spent upon them and experience has shown this expenditure to be one of those most advantageously spent. As most of the roads are well laid out and as there are plenty of them, there remains now (1863) mainly the keeping in repair of the roads to be attended to and not to build any new ones. Our endeavor now is, to do this at the minimum of cost. Statistics gathered on this subject, show good results and point out to us the means of arriving at still better ones. The present road law was made in 1810. That part of the old law which relates to the maintenance of roads is still in force, but that part requiring labor as a road-tax was abolished in 1831, and likewise most of the road police regulations. The appropriation for roads had to be increased 250,000 florins to pay for the abolished road-tax labor and to make up 170,000 florins previously received from tolls, which were also abolished in 1831. The system now is as follows: All town roads are taken care of by the towns. The State merely appoints and pays a road-master, so called, who superintends fifteen or twenty road-keepers and reports on the state of the roads, the reasons for their bad condition, if that be the case, what is needed, &c. The law for second class or county roads was formerly, that when they were of importance to several towns, they had all to help maintain the same. As this gave rise to continual bickering and quarrelling, in which the road suffered most, it was changed in 1856. They are now taken care of under the direction of the State and paid for partly by the State and partly by the towns in which they are situated. Most of the roads under this head are those which have risen in importance since the building of railroads, and are generally those that lie perpendicular to the direction of the railroad they are influenced by. The towns not having the means very often to properly improve and repair such, it was found necessary and expedient to give them the aid of the State, and in order to procure the necessary funds, all roads that run parallel to railroads and all those that had lost their importance by the construction of railroads, were in 1855 stricken from the list of state roads. These latter as the name implies, are wholly under the care and kept up at the expense of the State.

In 1835, the total length of the State roads was .	1,430.8	English	miles.
In 1855, " " " " was .	1,500.8	"	"
In 1855, by excluding several State roads, this last length was reduced to	1,142.4	"	"
In 1861, it had increased to	1,190.0	"	"

Second class roads, (keeping partly paid for by State.)

In 1835, the length of these was	467.6	English	miles.
In 1861, the length of these was	630.0	"	"

" So that the State had, in 1861, in all, 1,820 English miles of road to maintain, the towns helping to pay on six hundred and thirty miles thereof.*

* According to Chambers Encyclopædia, Baden has an area of about 5,900 square miles, and had a population, in 1858, of 1,335,952. It is probably this, or a little less, at the present time. Massachusetts has an area of about 7,800 square miles, and, according to the average of the computed populations in the supplementary tables of the census of 1865, it is, in 1870, 1,343,604.

Or, population per square mile in Baden, ... =226.43
" " " in Massachusetts, . . =172.26

By the same tables, accompanying the State census of 1865, we find that

"The statistics of the road repairs are kept in the following manner. The road-keepers are required to keep a record of all draught animals that pass in either direction. Horses that are being ridden, animals not before a vehicle, and teams going to and from the fields, are not counted. These records are kept only during the working hours. Likewise, not during the whole year, but only four months in each year, so selected as to give an average amount of travel. The travel on the road on Sundays and out of working hours is taken from a few observations; it is a very small percentage of the whole. At the end of the year these records and observations are collected and graphically represented on a map of the whole State. The different roads are drawn of a different thickness of line, according as the amount of travel on them is greater or less. The quantity of road metal used per yard of road, and the kind of metal used, give the data for another such map, in which the different colors of the roads represent the different materials used in their repair, and the figures on them and their thickness show the number of cubic yards per mile required to keep the road in order. Finally, we have a third map, which indicates, by the thickness of the several lines representing the roads and by the figures on them, the total cost per mile of repairing the road one year."

With this picture of a country happy and prosperous, in the possession of good and well-kept roads, it may be well to leave the subject.

Massachusetts wants, for her proper development, much better roads than she now has; and, reckoning for a period of say fifty years, she can have these good roads, and have them kept in order, at a less cost than that of keeping up the present poor ones for the same time. Besides this, we should see in the one case a healthy state of internal communications and trade; in the other an absence of both. Let each citizen so act and do his part, that these benefits may accrue to the Commonwealth.

the population per square mile of Massachusetts will equal that of Baden, above given, somewhere between 1890 and 1891. It exceeds that of Prussia, and probably equals that of France at the present day, both of which countries have systems of roads and road-repairing but little, if any, inferior to those of Baden.

APPENDIX.

CONTAINING SOME RECENT EXPRESSIONS OF PUBLIC OPINION ON THE SUBJECT OF BETTER ROADS AND ROAD MANAGEMENT.

[From the "Boston Daily Advertiser," April 25, 1869.]

Probably the heaviest tax paid by the people of Massachusetts, is that which they pay, in one form and another, for the privilege of maintaining some of the worst roads in existence. At this season, one may see anywhere in the country the full process of paying this tax. Indeed, within ten miles of this city, where we are in the habit of thinking that the arts of civilization are tolerably well understood, the traveller will find great county roads, like that leading through Somerville to Medford, for example, over which there is constant and important transportation, which are in such disgraceful condition that they might reasonably be the subject of indictment. Whoever complains will be informed that the trouble comes from the frost and from the season; but whoever takes pains to learn how far we of New England are behind other parts of the world—saving and excepting the rest of our own country—in the art of road making, knows that the real trouble is that we neither build our roads in the first place nor keep them in order afterwards; and that we thus continue to pay, directly and indirectly, a tax which, if we could have it assessed in the regular way, would cause a political revolution among us.

Undoubtedly we should be obliged to make a considerable original outlay, in order to secure the drainage of our roads and their construction of proper material. By drainage, we mean, of course, not the "crowning" of the road and the digging of a ditch for surface water, but the laying of proper drains under the road itself, so as to keep the ground free from the moisture with which the frost works its mischief; and by proper material we do not mean the solid filling of gravel or even loam with which matters are now annually made worse, but the broken stone, of various degrees of coarseness, with which a road-bed is made permeable by water below, and smooth on the surface. Certainly all this costs at the outset. But the road built as the road makers of the old world,—

and we are glad to say of one or two towns in this vicinity,—build them, scarcely shows the effect of "the frost coming out of the ground" in the spring, or of the early freezing in the fall; and in subsequent expense it is by far the cheapest. The traditional highway surveyor expects to find his roads badly cut up by the spring travel, and repairs the damage at a heavy cost to his neighbors; the regular road maker provides a highway which is never cut up while ordinary care is taken of it.

The experience of the one or two towns to which we have referred,—of which Waltham is the best example,—has shown that in the matter of annual expenditure, as a mere method of economy to the town treasury, it is cheaper to have roads well built, and then to keep them in high condition, than to undertake the annual repairs which are so familiar throughout New England. That is, it is cheaper to have a few men to watch for the beginnings of any wear upon the roads, and to mend a defect when it first appears, than it is to wait until the trouble becomes serious, and then set a large force at work. One man, with a shovelful of broken stone, can prevent what it may require half a dozen, with a team, to cure after a few months' neglect.

But the heaviest part of our highway tax is no doubt that which is levied upon us by the destruction of horseflesh, the impeding of public travel, the wear of vehicles, and the increased cost of transportation over our poor roads. This is a tax which, without any assessor or collector, is inexorably exacted from every barrel of flour, every bag of grain, every box of goods, and every person, passing over our travelled roads. As it is levied indirectly, and more than all, as we have always paid it, nobody thinks of it. But it is one of the heaviest burdens resting on the people of Massachusetts and of New England, borne by those who rank among the most thrifty and progressive people on earth, and who, nevertheless, in this every-day matter, are demonstrably rather more than two thousand years behind the world.

EXTRACTS

From a Report of the Committee on the Appointment of Road Engineer, to the Town Meeting in Newton, held May 3, 1869.

Your Committee consider that the question of the employment of a road engineer largely involves the question whether the roads of the town of Newton shall continue to be repaired in the old-fashioned, temporizing manner, or whether the work shall be done thoroughly. And they are satisfied that the true interests of the town will be best promoted by employing the best attainable engineering skill; by having the roads over which heavy teams travel macadamized with broken stone, and also by having the gravelled roads, as well as the broken stone roads, underdrained in all places where springs manifest themselves, or where higher land adjacent to the road makes it wet and muddy. The test of a road is its condition in bad weather and its power of sustaining heavy teams without being cut up in ruts or pounded full of holes; and certainly some of the roads in Newton, although in fair condition in dry weather are not able to resist rains nor sustain heavy teams. Should the town purchase a stone-crusher, and decide to repair with broken stone, it will still be no less necessary to underdrain, or, in some suitable way, relieve the road-bed of water, or the macadamizing will not be effectual. A good engineer has the gathered and recorded experience of communities and nations. He knows, or should know, the relative qualities and degree of adaptation of different kinds of rock for the purpose of macadamizing; he can detect anything in the subsoil which renders it unfit for a road-bed, and suggest a remedy. He can calculate the area of rainfall in a given precinct and determine the amount of drainage, and the necessary size of culverts and bridges. He can grade drains, both surface and concealed, at the proper angle, and select the most feasible route for their construction. He would know about different crushing machines, their prices and relative value. His knowledge and suggestions would save the town many times his salary by enabling the town to avoid useless and perhaps costly expense. And it may be hoped that in a couple of years our present superintendent, or some other person, would have thus received such hints and information as would enable him to manage our roads without the supervision of an engineer.

The town of Waltham macadamizes its principal streets. It keeps ten to fifteen men under constant employ, and much of the time on the roads; one of these men has thus been engaged for twenty years. Their superintendent of roads has held his situation twelve years; his salary as road master is $800; he is also paid $100 as highway surveyor. Waltham, in 1865, had fifty-one miles of road, and for the previous seven years its roads had cost the town an average of $3,357 a year, or about $66 a mile. It cost that town the past year for repairs of roads and clearing off snow $6,000 for sixty miles of road. The town of Newton expended for repairs and clearing of snow in the same time on eighty-two miles the sum of $14,523 or $176 a mile. It will thus be seen that our system of partial or incomplete repair is almost twice as expensive.

But this is taking the most narrow view of the subject. An editorial in a recent Boston paper says: "The heaviest part of our highway tax is no doubt that which is levied upon us by the destruction of horseflesh, the impeding of public travel, the wear of vehicles, and the increased cost of transportation over our poor roads. This is a tax which without any assessor or collector is inexorably exacted from every barrel of flour, every bag of grain, every box of goods, and every person, passing over travelled roads. As it is levied indirectly, and more than all as we have always paid it, nobody thinks of it. But it is one of the heaviest burdens resting on the people of Massachusetts and of New England, borne by those who rank among the most thrifty and progressive people on earth, and who nevertheless in this every-day matter are demonstrably rather more than two thousand years behind the world." That this is not imaginary, statistics show. General Morin found, by careful experiment, that "carriages on springs, drawn upon a new road covered with gravel five inches thick, required in tractive force *one-eighth* the load; upon a solid causeway of earth, with gravel one and a half inch thick, one-tenth the load; upon a causeway of earth in very good condition, one-twenty-sixth the load; upon a broken stone road, very smooth, one-forty-fifth the load; upon a broken stone road, moist or dusty, one-thirtieth the load; upon a broken stone road, with ruts and mud, one-twentieth the load; upon a broken stone road, with deep ruts and thick mud, the tractive force required was one-tenth the load." It will thus be seen that the smooth causeway of earth, in very good condition, required but one-twenty-sixth of its weight in tractive force to draw it, while the smooth broken stone road required only one-forty-fifth the weight of the load in tractive force.

Before the invention of railroads the attention of engineers in Great Britain and on the continent of Europe was largely given to the construction of common roads and turnpikes, upon which the heavy travel of those nations depended. Telford built, more than fifty years ago, 800 miles of road in the highlands of Scotland, still admirable and in constant use. The roads built by the French engineers of the first empire are still among the finest in the world. No one that has travelled over the Simplon road, or the Mt. Cenis road, or from the St. Bernard pass to Martigny, or all around the Bay of Naples to Sorrento, on a road built by Murat and which the lazy Italians have had the grace to let alone, can possibly resist the claims of a good, smooth, hard road. It adds new charms to scenery and imparts a fresh zest of life. But to come back again to statistics. MacNeill constructed a machine to test the amount of tractive power required on different roads. This was carefully tested by many very eminent engineers. Their experiments showed, uniformly, that the force of traction is, in every case, nearly in exact proportion to the strength and hardness of a road. They found that on a road made with a thick coating of gravel, a load, which required the power of one hundred and forty-seven pounds to draw it, could be drawn on a broken stone road with a power of sixty-five pounds; and on a road of broken stone of great hardness, laid on a foundation of large stones set in the form of a pavement, the power required to remove the same load was but forty-six pounds; and this last road effectually resisted frosts.

On the 6th inst., your Committee visited Waltham, and found the broken stone road there dry and hard. It will sustain loads of six tons without being cut into ruts. Returning, we came down through Waltham street, and, observing the instant of passing from town to town in the changed character of the road, we passed on to Newtonville over our fine old avenue! The first road required scarce any mending. The last was cut up with ruts and full of mud, and workmen were dumping gravel four to six inches deep upon it. On the Waltham road it required not more than one-fortieth the weight of the load (say 20 lbs.,) in tractive force to draw it, while on the main road of the good, rich old town of Newton, it would have required one-eighth the weight of the load (100 lbs.,) in tractive force to draw it. That this criticism is not especially in the interest of persons driving in light carriages and for pleasure, may be seen from the fact,—as demonstrated by careful and extended experiment,—that resistance to the onward motion of the carriage or cart, arising from roughness of the road, is always in proportion to the weight of the carriage. A double weight will offer double resistance, and a triple weight triple resistance, and so on. In truth the principal objection of some, who have had little or no experience with the broken stone road, is, that it is unfit to drive horses rapidly upon; such a road improperly made, or imperfectly hardened, is indeed unsuitable for rapid driving; but it requires only a journey of three hundred miles* to see millionaires driving the finest trotters in the world, upon roads as smooth as a floor, made entirely of stone.

Can any measure be more likely to increase the popularity of our town, or add to its population, than to construct roads as solid, smooth and perfect as the nature of the case will admit?

* To the Central Park, New York.

[Correspondence of "Advertiser."]

It is recorded of the Rev. Mather Byles, that he saw one day a chaise, containing the town clerk and a selectman of Boston, stuck, before his house in School Street, in a quagmire which he had long tried to induce the town to mend. He pulled off his hat and cried to them that he was heartily glad at last to see the authorities "stirring in the matter." The writer is as glad as the reverend doctor was, to see so good an authority as the "Advertiser" stirring in the matter of our *roads*, though by no means in so muddy a style as the worthies referred to. A month or more ago you published extracts from the report of the committee appointed by the town of Newton, recommending the employment of "a road engineer, to have supervision of the roads in connection with the selectmen, and to survey the town with reference to laying out new streets." This report was adopted by the town. A number of your readers in our beautiful suburban towns wish that you could find space to publish the whole report, or that the selectmen of Newton would circulate it as a tract throughout the environs of Boston, where our excellent authorities are contented to believe that roads, which are well nigh impassable during seven months of the year, are good roads, because they are in fair order through the summer months.

It would seem that in such rich towns as those which surround Boston, the public had a right to demand as durable, solid and smooth highways as lead into any city in the world. But we are told that they have worse roads in Little Pedlington, or the frosts in New England are so severe, or there are horse railway tracks in our streets, or we have such heavy travel over *our* roads, that we ought to be proud of the roads instead of grumbling at them. So the "seelic-men" continue to dump big rocks and loam, turf and brickbats into the streets; and if the town owns a "stone-cracker," the sharp fragments are strewn along the ways, to be beaten down by hoof and wheel, as if a layer of gravel and a roller existed only in the heated dreams of condemned highwaymen. Next, the autumn rains, the winter frosts and the spring winds do their work, and the expressmen, and the farmers, and the doctors, and the teamsters, and the milkmen, try to do theirs, over the torrent-beds and springholes we fondly call our "streets," ruining their wagons, killing their horses, and endangering their souls by consequent profanity.

It is not those who drive for pleasure who are concerned in this matter; it is the crowd of hard-working men we have just named. Such men, if they could read the report referred to, would see that "the heaviest part of our highway tax is no doubt that which is levied upon us by the destruction of horseflesh, the impeding of public travel, the wear of vehicles, and the increased cost of transportation over our poor roads." That, while over a bad road "there is required in tractive force *one-eighth* the load," over a properly made one is "required only one forty-fifth the weight of the load in tractive force." "If we encourage with good roads, we shall speedily derive therefrom a great and permanent advantage." "Can any measure be more likely to increase the popularity of our town, or add to its population, than to construct roads as solid, smooth and perfect as the nature of the case will admit?"

But space will not allow us to quote the demonstrations of the report. The public will have durable and smooth roads when it has been pointed out that roads which are cut up and undermined for more than half the year are *bad* roads; that what is spent in making good roads, under the superintendence of an accomplished civil engineer, is saved ten times over in horseflesh and carriages; and that by proper building of roadways, by under-draining and macadamizing and rolling, we can have roads worthy of the name even in frosty New England. The irrepressible Yankee builds the best carriages, and breeds superior horses: what a pity it is he cannot make roads for them to travel on that will not wrench the former apart before the gloss is off the panels, or lame the latter by springholes, ruts and stones. The fact stands, however, that Americans are hehind all civilized nations in the art of road making. Will you continue to call the attention of your readers to this melancholy but true statement, in order that you may share the proud eminence occupied by the great Highland road maker, of whom it is sung:—

"Had you seen these roads before they were made,
You would hold up your hands and bless Marshall Wade?"

SUBURBANUS.

SECOND PRIZE ESSAY

ON THE

CONSTRUCTION AND REPAIR OF ROADS.

By Samuel F. Miller, C. E.,
PROF. OF MATHEMATICS AND FARM ENGINEERING, MASS. AGR. COLLEGE.

A "treatise on the science of road making, and the best methods of superintending the construction and repair of public roads in this Commonwealth," being called for by a Resolve of the last legislature, I submit these few suggestions, not as exhaustive of the subject,—for my time will not allow me to attempt such an essay as seems to have been anticipated by the Act of the legislature,—but as the results chiefly of my own observation and experience.

The first roads of a country usually follow the high lands and ridges, when the intervening valleys are narrow, or the land is wet or heavily timbered,—and military roads are not an exception,—for the obvious reasons of greater security, less water to contend with, and being more easily built. Besides, roads built by citizens are at first only improved foot or bridle-paths leading from house to house; and the earlier settlers live upon the hills for good sanitary reasons. Hence, the problem of changing the location of the old highways is sure to come.

An Indian trail is a better route to follow for a road than a white man's path, for those great travellers, the Indians, learn intuitively to economize their strength. But our farmers are slow to learn that it would pay them back fourfold to make better roads, for many of them might thus be cut off from the travelled way, and they are not in the habit of figuring closely on the value of time, the cost of "wear and tear," animal strength and feed. They would not be greatly moved by the statement that their roads to "town" might be made, within a reasonable cost, so that they could haul three times as much as they now do at each load, and make the trip in one-half the time. They are very conservative, and greatly averse to having their fields cut up again, especially in any other way than "square across."

And it must be admitted that it is a serious matter to take the travel from the doors of citizens, even though the public are obliged to provide an outlet, or to change the road so as to seriously impair the frontage of their buildings. And, on the score of public expense, we would not advise this, except it should be justified by the wants of travel.

But whether old roads are to be altered or new ones to be laid, the principles of location and construction are substantially the same. We will consider the whole subject under the heads of *Location*, *Grades*, *Road-Bed*, *Drainage*, *Bridges and Culverts*, *Repairs*, and *Superintendence*.

The science and art of road making have been carried to great perfection in England and France, and we have many books and statistics on the subject from these sources, as those of Tredgold, Wood, Parnell, Morin, McAdam, Telford, Rankine, and a host of others; while in this country, such manuals as those of Gillespie and Mahan, giving a digest of foreign experience adapted to American wants, are eminently simple, practical and suggestive; so that there seems to be nothing lacking except to notice the more important points as bearing upon the necessities of our own State. It is believed that skilful men to do the work will make up all the deficiencies of the books, and without them the books will be of little use.

LOCATION.

When a road is to be located, there are termini that are fixed, and generally one or more points between. As a matter of economy, there should be no deviation from a straight line between fixed points without sufficient reason. But it should be borne in mind that road distance is measured on the surface, and not horizontally, as in land surveying; and the old illustration is to the point, that "a kettle bail is no longer when lying upon the rim than when erect."

Where a line *pretends* to be straight let it be *perfectly* so, and let the centre line of the road-bed, and side ditches, show this when completed. Then in place of angles there should be curves of definite radii, properly fitting the ground; for these shorten the distance, and are far more graceful, and more pleasant to drive over. The principal exceptions to this rule would be at cross roads, and perhaps in villages. While the tangents may be described by needle bearings and distances, the curves should be designated as to the right, or left, with such a radius, and such a distance. Stones, three feet deep, should be set at the tangent points, or in the fence lines opposite, and then the line is permanently fixed, and may be tested upon the ground at any time by the aid of a description as above.

Where the stones are set and preserved the record of the bearings of the needle is of little use, but otherwise it is of great value, provided you know whether the true or the magnetic meridian was meant, and, if the latter, what the variation was at that date. The difficulty in the matter is,—and our ordinary land surveying suffers still more,—most of our own surveyors do not know what the local variation is, or how to obtain it accurately; and in fact it is a problem of considerable difficulty with ordinary means.

Hence arises the great uncertainty in retracing many old surveys where the corners are gone, and troublesome lawsuits have arisen; for the needle in this State has never, so far as known, pointed within five degrees of north; and it is constantly changing at irregular rates, varying from one to six months each year: and the extreme difference of variation at the same time in different parts of the State is about three degrees. So that the record of one place may not answer for another ten miles distant.

In view of these facts it may not seem out of place here to state a very simple remedy. Let each town, or district of several towns, be provided with a true meridian in its central part, marked by three stones set securely in the ground two or three hundred feet apart, and let a yearly record be kept of the magnetic variation at each place. Then let every surveyor be required to test his instrument yearly, and every survey that goes upon record have appended a properly sworn certificate, stating whether the bearings are true or magnetic, and what the surveyor calls the variation at that date. The expense of fixing these points, with sufficient accuracy for all surveying purposes, would be trifling compared with the immense advantage gained.

Physical reasons chiefly, such as cost of construction and maintenance, and proper grades, should cause deviation from a straight line, but there are, in extreme cases, considerations of "land damage" that may have weight; but a most careful balancing of first cost, with permanent detriment to the line, should be made.

In locating a road, the travel it is designed to accommodate, the character of the ground to be passed over, and the money that can be raised, are supposed to have been carefully considered in order to fix very nearly the maximum grade that can be allowed; and then the line must be made as direct as the nature and surface of the ground, and the passage of streams, will permit; always taking into account subsequent cost of repairs, as well as first cost.

In order to do this properly levels must be taken throughout the line, and a profile made, to show to the eye on paper the grades that can be obtained, with the "cuts" and "fills" at every point; and this, as finally established, should form a part of the record. Then, in construction, let the cuttings and fillings be plainly marked upon stakes set once in fifty or a hundred feet to guide the workmen. This grading by the eye is expensive in the end, for materials are apt to be taken from, and carried, to the wrong places.

It is a mistake, as often stated, that the cuts and fills should balance each other; on the contrary the latter should be usually large in excess, because casting, scraping, and wheelbarrow work, are so much cheaper than hauling, the road-bed is made so much dryer by elevation, and proper drainage easier to obtain.

It is best to adopt such a route as *ought* to be built, on a line of importance, even if at first unable to carry out fully the plan, for, with a definite recorded plan to work to, the completion may occupy several years, while, in the meantime, a good passable road can be maintained. This is the American plan of building railways, which has won for itself so high English commendation that the East India Company has been considering the question of employing our engineers to build railways in India.

The width taken for "right of way" should be designated at

every point on the plans of record. These should never be less than fifty feet, and greater widths must be taken at some points of heavy cutting or embankment; and, in general, there should be width enough for all purposes of drainage, and preservation of the road-bed, and for obtaining materials for construction and repairs; except for ballasting, which must often be obtained from gravel beds outside. It is better to have too much than too little.

GRADES.

There are two general classes of carriages to pass over roads, one for freight, and one for passengers. The former wish to carry the heaviest loads, and the latter to make the best time. Undoubtedly where heavy loads preponderate, there is greater need of light grades, but in general, the road that is best for the one, is the best for the other class of travel.

We will designate grades in the usual way by a fraction having unity for the numerator, and the distance required to rise one foot for the denominator; for the simple reason that this fraction expresses the exact ratio of the draught to the load, as it would be were there no friction. For instance, a grade of one foot in one hundred is called $\frac{1}{100}$, three feet in one hundred $\frac{1}{33}$, and five feet in one hundred $\frac{1}{20}$, etc.

It seems to be well established that undulating grades of $\frac{1}{100}$ are no detriment, but, on the contrary, are rather beneficial, as giving better longitudinal drainage.

It is evident, too, that the grade should never exceed the "angle of repose," if possible; that is, the horse should not be obliged to hold back his load in descent, but the friction should be sufficient to counteract the force of gravity. This "angle of repose," or limit of grade, for different road-beds will vary from $\frac{1}{50}$ to $\frac{1}{10}$ on ordinary roads. It is easily ascertained from the draught upon a level with the same character of surface, for this is just equal to the friction, which would be the same upon an incline. For instance, if the draught upon a level, or what is required to overcome friction, is $\frac{1}{20}$ of load, then this fraction would exactly express the maximum grade to be allowed for same kind of surface, since, as stated above, the same fraction gives the ratio of draught to load to overcome gravity alone. So the "angle of repose" is the grade expressed by the fraction which indicates the ratio of draught to load upon the same road-bed on a level.

Very many experiments have been made in Europe to find this ratio upon every variety of surface, and we have full tables of the results. We will only state briefly that upon such roads as we shall have in Massachusetts, varying from those of earth and gravel mixed, to those of broken stone, in good order, according to these data, the draught would range from $\frac{1}{10}$ to $\frac{1}{50}$ of load on a level. Now, *theoretically*, to find the power required upon any given grade, we have only to add to this the ratio of perpendicular rise to length of surface, i. e., the fraction expressing the grade. For instance, if the power on a level, with the same road-bed, is $\frac{1}{30}$ of load, on a grade of $\frac{1}{20}$ the power required would be $\frac{1}{30}+\frac{1}{20}=\frac{1}{12}$ of load. But, *practically*, this calculation for power required on grades from that on a level would lead to wrong conclusions, for a horse does not pull at so great advantage on an incline, on account of the position of his body, not being able like a man, to throw the centre of gravity forward at his will, and not having so good foothold, and the load, too, being thrown more upon the hind wheels, there would be increase of friction on a yielding surface. So we must resort to experiment again, as upon a level; and the general results are, that a horse can haul about $\frac{4}{5}$ as much on a grade of $\frac{1}{50}$ as on a level; $\frac{2}{3}$ as much on a grade of $\frac{1}{33}$; $\frac{2}{5}$ on a grade of $\frac{1}{20}$; and only $\frac{1}{4}$ as much on a grade of $\frac{1}{10}$ as on a level.

These are condensed general statements from English and French experiments; but the State should at once institute full experiments upon her own roads, with the vehicles in common use, and with such as shall prove the best.

Now the maximum grade upon any line being fixed from such considerations as have been named, it still remains a question of great importance how much you will increase the distance, or the cost, to reduce grades that are within the prescribed limits. No special rules for each case can be laid down, but they are questions for the engineer on the ground, in view of all the facts and principles that should control the decision. We will mention a few *general* rules that are well established.

1. The capacity of a line of road should not be limited by a portion which is a small percentage of the whole. If possible, let the greatest load that can be drawn up the steepest part be a

fair one for the balance. It would be justifiable to make considerable extra expense to reduce two or three hard points; certainly, if in no other way, by making the road-bed there so much better than the rest as to compensate for the extra steepness.

2. Where the difference in levels between two points on your line is large in proportion to the distance, avoid, if you can, any *counter* grades, as so much loss in your ascent. Within certain limits, *level* portions interspersed with varying ascents, always supposing that the line fits the ground, are no damage, but they serve to refresh the horse in a small degree.

3. Do not go *over* a hill if you can help it. It is usually no further around it, and if it is, the greater distance and cost may be amply compensated by the saving in time, wear and tear, and animal strength. Gillespie says it would be better to lengthen a road twenty times the perpendicular rise saved than to go over the hill; which is to be understood as a popular, rather than a scientific statement. When you *must* go over, the principles of the second rule would apply to reach the summit on either side.

4. It will not pay to make much more distance, or cost, to reduce grades not exceeding $\frac{1}{33}$ that are not very long. This is a general statement, to which there might be exceptions on some very heavy freight roads. On lines of such a maximum driving carriages will not usually change their average speed for level except to go a little faster down, and a little slower up, the steepest parts. But grades of $\frac{1}{25}$ and upwards may fairly be considered as worth reducing, according to their respective detriments. Gillespie, under the head of "Profits of Improvements," illustrates a good method of figuring such a problem.

It is suggested that our public roads might properly be divided into two classes, according to facts of traffic existing, or what may be expected from the greater facilities proposed; and the highest grade to be allowed in the first $\frac{1}{20}$, and in the second $\frac{1}{10}$. First class roads then would have no grades up which, with a light carriage, a fresh team might not trot at moderate speed, and down which the common speed would be that of the average on a level. And we submit the question whether all the roads of the State might not in time be brought within the limits of the *second* class.

Road-Bed.

Quite equal in importance to the grade is the surface to be driven over.

Theoretically this should be unyielding, smooth, and impervious to water. It should be a roof, shedding the water each way from the centre, from eighteen to twenty feet wide, except in or near villages, where it may be thirty feet, or even fifty. An elevation of six inches at the centre is required, and the lateral slopes should be *planes*, rather than curves as is usual, that the water may be carried off more quickly. We mean by the "road-bed" simply the drive-way, and this should be as narrow as perfect safety and convenience will permit, on the score of economy, both of construction and maintenance. Bars across the road for turning the water into the ditches are only necessary on the heavy grades of inferior surface, or of such roads as are not kept in good repair; and when needed, should be built V shape, with the angle up the grade, if practicable, that they may strike both wheels at once. An exception to this rule would occur when it is desirable to turn all the surface water to one side; and in that case the slope of the surface would fall one foot in the width of road-bed to the same side. Such a shape of road-bed is desirable in very short turns, as in zigzag roads up steep mountains, and around sharp angles of hills, where the width of driving surface should be a few feet greater, and the slope inward.

The materials of the road-bed must be the best that can be afforded to approximate as nearly as possible to the true theory of a perfect road surface.

The importance of this is evident from the facts established by experiments, that from two to three times as much can be hauled on a broken stone road as on one of gravel, both being in equally good order; and from four to five times as much upon a good pavement of rectangular blocks of stone.

The broken stone road, and the block pavement, have been so long in use, both in this country and in Europe, and in fact throughout the civilized world, and their modes of construction and repair so fully discussed in treatises on road making, even to the minutest details, that it would be superfluous to dwell upon the subject here. It may be remarked however that the Nicolson pavement of wood, or its essential principle which has been tried in this country now for ten years, bids fair to largely supersede the stone blocks. It meets one great desideratum, among others, of furnishing a less noisy, and more elastic surface than stone.

But most of the roads of the State must be built of loam, clay, gravel, and sand, more or less mixed, and with these materials we must do the best we can.

A road-bed of clay is good in dry weather, and one of sand is tolerable in a wet time, and the latter is but slightly affected by the frost, but we need one which will be good at all seasons.

Not being able to make the surface " water tight," there must be a substratum that will carry the water from above, and that from below, as quickly as possible to the drains. Whatever may be used for this under layer, pebbles greater than one inch diameter should not be placed within ten inches of the surface, for otherwise, when there is a compact foundation, they are sure to work up, from the downward pressure of gravel, forcing itself beneath them. Clean gravel, rejecting stones above an inch diameter, put on in successive layers of three or four inches, to the depth of ten inches, each layer being well worked down with a heavy roller, assisted by the daily travel, with a small admixture of finer gravel or loam upon the top, in some cases will make as good a surface as we can expect for the majority of our best roads.

Pebbles, about an inch in diameter to two or three inches, placed upon the surface, will not pack together, on account of their hardness and smooth, round surface. It is well understood that McAdam and Telford roads are made of *broken* stone —*angular* fragments.

A pure sand road is improved by a topping of clay or loam ; and it may be six inches in thickness, or more, according to the depth of the sand.

Quite passable roads have been made in a region where a vegetable mould and loam lay upon pure clay, and no sand or gravel was at hand, by leaving the surface as nearly as possible in its natural state, and flanking with deep ditches, throwing the excavation from them *out*, and not upon the road-bed.

A very good use of stones, if they are abundant, is to place them in a slough that must be crossed. But if it is deep, a better method is to use poles or brush, in alternate layers, first across and then lengthwise, covering the whole with not less than two feet of the best material at hand. If a layer of two or three inches even of coarse gravel can be put upon an earthen road once a year, it is a great improvement.

Sidewalks should be from four to six feet wide, sloping inwards with a fall of two inches, the top to the depth of six inches being fine and clean gravel, with a very porous substratum like that of drive-way; there being no danger in this case of forcing up the large pebbles from beneath.

The natural place for sidewalks in villages, and wherever they extend into the country with light road cutting and filling, is by the side of the fences; but otherwise thay must be at the side of the drive, and form a part of the road-bed, to lessen the expense of construction. In either case the walk should be a little higher than the centre of the drive; and in the latter, on *gravel* roads, the best way is to give an extra width of road-bed, and lay planks lengthwise upon 4″ × 6″ cross pieces, thus giving free passage for the surface water at every point. But away from the vicinity of villages the expense of sidewalks will usually be saved, and the footman will take the drive-way.

Embankments must have a greater width where railings are needed, and these should always be placed where the fill exceeds three feet. The clear width between the railings should not be less than twenty-two feet, and three to five more in case a sidewalk is required, and two feet additional on each side for the strength of the railings. The following cross-section shows the plan upon embankments, with a plank walk on one side, and railings, at minimum width:—

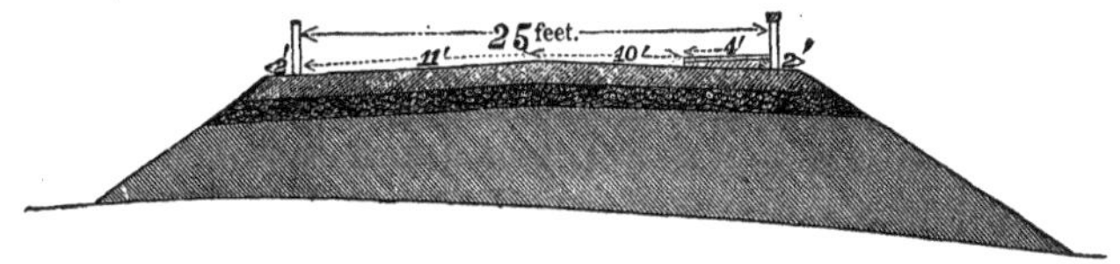

Drainage.

This part of road-building is really the most important of all. A thoroughly underdrained road-surface will not be seriously affected by the heavings of frost, will dry up speedily after rains, and will be passable at all seasons of the year by the average loads, though the materials composing it may be of an inferior character ; *i. e.*, more depends upon the *drainage* than upon the materials of surface.

The first thing is to obtain a *substratum* which will drain the surface like a sieve, either into tile drains beneath, or into open ditches at the sides, which should be three feet below the crown of the road, and have a fall not less than eight inches to the hundred feet. This cannot be done thoroughly with *less* than one foot of clean coarse gravel, or pebbles, rejecting all of an inch diameter and less. The foundation for this should be graded with a crown corresponding to the top, that the water reaching it may pass off the more rapidly. The difficulty is that this substratum becomes clogged, and the drainage proceeds too slowly ; or the side ditches become filled up, and the surface is not drained at all.

The best remedy for this is a line of tiles, well laid directly under the centre of the road, lengthwise, three feet deep, communicating with the side ditches where there is a sufficient fill, with the ends properly protected from washing out, or filling up. These should not be needed on banks of two feet and more, but they are especially useful where the road-bed is near the natural surface, or in cuts ; and then the side ditches need not be more than half so deep as otherwise.

Most cuttings have a wet and dry side, and in such cases it would be better to lay the tiles nearer the wet side. The expense of this would not often exceed $1.25 per rod, and in many cases it would be somewhat less ; and there would be a large saving in the digging, and keeping open, of side ditches. There is no doubt that, in very wet places, the tiles would be the cheapest in the end, and they would make by far the best road.

The side ditches should have from one to two feet width on the bottom, the slopes on the inside not greater than two to one, and on the outside an average of one and one-half to one.

The following cross-section shows a general plan for road-bed and side ditches in cuts, and on fills not exceeding three feet. Of course, as the bank rises, the ditches become less in depth.

A great saving of water at the side is made by a ditch above the slope of a cut on the upper side, taking the surface-water to the side of the next fill. The dimension of these surface ditches will depend upon the amount of water to be provided for, and their depth may be diminished by depositing the earth taken out on the lower side next to the top of the slope.

The following cross-section shows the plan of a cut, with width of road-bed at twenty feet.

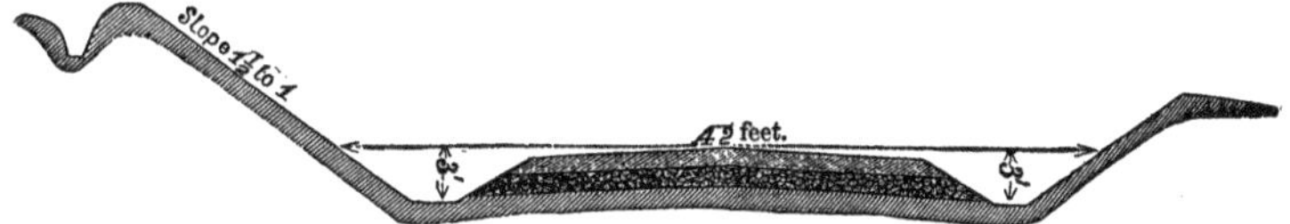

Cross-section, of same cut as above, without a substratum of pebbles, and a line of tiles in the centre, showing a saving in width of cutting and depth of side ditches.

Where no gravel or pebbles can be obtained, the under-draining will undoubtedly make the best road at the least cost.

In cities and villages it is desirable to be able to drive to the edge of the sidewalk; and this will not allow a depression for surface-drainage *exceeding* one foot below the crown, depending upon the width of driveway, and in many cases it should not exceed six inches. Hence, in such cases, a system of under-drainage becomes indispensable, unless there is a very deep substratum of clean gravel or sand.

An increased width of road-bed may make *two* lines of tile-drains necessary, one towards each side; and even in the ordinary width of twenty feet it may be expedient in some very wet places.

On the other hand, the under layer of pebbles will not generally be necessary on banks of three feet and upwards, though it will diminish the heavings of frost, and make a dryer road in the spring.

In side-hill cuts, where much water comes from the upper side, culverts beneath the road should be placed at short intervals to relieve the hillside ditches. All the ditches named should communicate as quickly as possible with natural water-courses, and special care be taken in their construction, that they may not be washed or overtaxed in a freshet.

In general, no work upon a road pays better than that which tends to secure a perfect system of drainage, keeping the water level three feet below the surface of the road. And it may be observed that the vicinity of the road should be kept clear of trees and bushes, as they obstruct the sun and wind and prevent evaporation. Hence shade-trees, set along the margins for beauty and comfort to the traveller, should not be placed too frequently, and only in dry and exposed places.

Bridges and Culverts.

Upon this subject abundant instruction has been published in a multitude of works, both in a scientific and in a popular way, and we will briefly state a few general principles.

The leading point in these structures is to make them *permanent.* Ample provision for the severest freshets—like the one last October—must be made by a careful and exhaustive study of each locality. Above all, secure for these costly structures foundations that cannot possibly settle beneath any pressure they may have to bear, and which cannot be undermined. The safest way is to place them very deep, and heavily timber the bottom if it is at all yielding or treacherous in its nature. Quicksand, the most fickle of all, will support the heaviest of structures if it can be confined. Piling, topped with a platform of timbers alternately crossing, or the tops embedded in a thick layer of concrete, has never failed the engineer when the work has been skilfully done, and properly protected by rip-rap or otherwise.

In cases where a soft material lies upon rock, not many feet below, the excavation must extend to the rock-bed, and this is the best kind of a foundation.

Special attention should be given to the lower ends of culverts to prevent their undermining. This may be effectually done, if necessary, by transverse walls of masonry, or rows of sheet piling, placed beneath the bed of the stream, and the intervals filled with large stones.

A great thing is to give an *abundance of water way.* By

compressing and raising water in its flow, its destructive power is largely increased. Comparatively weak structures will often outlive stronger ones simply by this wise foresight.

It seems more important that these works should be thoroughly built upon highways than upon railways, for the latter usually have more competent supervision after completion.

For the same reason, the superstructure of bridges should be of the simplest kind—that which will need the least care and be most easily repaired. There is no difficulty in making firm spans of twenty-four feet with stringers alone, without braces or truss, provided good timber of that length, fourteen inches deep, can be obtained.

For spans of forty feet and upwards, the "Howe Truss" stands preëminent among wooden bridges; and there are styles of iron bridges, for spans of thirty feet and upwards, which may be the cheapest in the end, if the first expense can be borne. But since there is no danger of fire, we are inclined to prefer the wooden bridge generally.

We will only add that on all bridges of a considerable span, signs, forbidding fast driving, should not only be put up, but, what is unusual, they should be *enforced;* for the vibration caused by rapid driving is tenfold more injurious than the same loads at a walk.

Repairs.

A great deal of money is washed away on our highways for want of constant supervision and prompt repairs. A côstly culvert or bridge may be saved by observing the action of the water in time, and applying the necessary remedy.

There is constantly forming in the centre of the drive a horse path, and at the sides ruts, both of which prevent surface-drainage to the sides, and form excellent channels on grades for the water in a rain-storm. Water stands in these at different points, the surface is softened, and they are deepened more and more rapidly; for the blow of the wheel increases with the distance fallen. A great rain-storm comes, the road is gullied, hundreds of yards of good ballast are wasted, and the charge is made to the "dispensation of Providence."

Let the road be divided into convenient sections, and a man constantly employed upon each, with cart, scraper and roller, to fill up the depressions as soon as formed, keep the ditches open, sharply watch the bridges and culverts, and keep the whole in perfect order. After rains, let him drag his scraper, made of heavy timbers, V shape and shod with iron, over the road-bed to fill up the ruts, cart on the deficiencies in material, and follow with the roller, throwing out the large and loose stones. Thus the periodical repairs would be comparatively small affairs; and, with no more expense, we should have far better roads, and of *constant* excellence. *Now*, a season of repairs is much dreaded by travellers, and it often takes several weeks to get the roads into as good a passing condition as before. One man, by the aid of a team and the implements named, can "keep up" from three to five miles most of the time, provided it is first well constructed.

When general repairs are made, which should be in the spring, and again in the fall, if needed, let the old surface be disturbed as little as possible, as this is firmer than new materials, and let the best that can be obtained replenish the waste. The customary way of scraping out the ditches, and placing the wash of the roads and adjoining slopes in the centre of the drive, is generally the poorest kind of road repairing, for it is loam and sand chiefly that is washed into the ditches. The fresh deposits on the road-bed should be selected, and placed, and worked down, with the same care and labor as of first construction, that the repairs may not for any time be a detriment to the travel.

Superintendence.

It is a cardinal principle that all work should be managed by experience and skill; and this rule is seldom violated by men of intelligence in the conduct of their own affairs.

In Massachusetts, on her fifteen hundred miles of railway, we are told yearly how many passengers and tons of freight are carried one mile, and the average cost of each per mile; but we are left to *guess* at the amount and cost of highway traffic, though it is not much, if any, inferior in amount, and in which the whole population are directly and personally active partners. When this whole business can be reduced to a science, like that of railways, and the results given in tables with approximate accuracy, showing the cost of building and repairing yearly done for public travel, and the amount and cost of transportation on our highways per item, then we can tell how much money should be expended in this way, and how to use it to the best advantage.

When we consider the great expenditure of the people for public transportation, not only in road taxes, but in time, expense of vehicles, wear, and animal strength, an amount far exceeding railway expenditure, and one borne more directly by all, it certainly seems to be a matter worthy of legislative investigation and supervision, ranking in importance financially above any other public interest in the State.

To constitute an efficient superintendence of this whole interest, making a complete system, operating uniformly throughout the State, and founded upon the most enlightened science and practice of the age, it is suggested that there should be a board of highway commissioners for the State, who shall be appointed simply for their fitness for such a charge, above any political considerations, whose duty shall be, by the aid of a competent corps of engineers, to take charge of the laying out, building, and repairing, of all the public roads in the State; to fix upon the various plans and principles by which all this work shall be done; gather facts of traffic upon the different roads, and the cost of the same upon those of different characteristics as to grades and road-bed; make re-surveys of old ways, with maps and profiles, to show what they are, and how they may be improved; revise old fence bounds and place permanent land marks; perfect county records of all its highways, that everything important may be shown there, both to the eye, and in accurate description; and make experiments upon different grades and surfaces with faithfully recording dynamometers, to ascertain with practical accuracy the draught due to grade and to road-bed, at various speeds, and with vehicles of varied build, four-wheeled and two-wheeled, single and double, with wheels of different diameters and breadth of tires, with and without springs, and in short, under such a variety of conditions, both as to road and to carriage, as to enable the public to advance towards perfection in the art of highway traffic.

It is not suggested to modify materially the present mode of raising and appropriating money for highway purposes, but only to provide a faithful and competent supervision of the expenditure. The plans of the engineers must be approved by the county commissioners, or the towns, as now, before they can be carried out. It would seem, however, that existing laws need to be changed in *two* particulars at least to carry out such a plan as has been suggested. In the place of a jury, in cases of non-agreement between parties, who have power, not only to revise the award of damages, but even to change the plan of the location of a road, the superior court should appoint a commission of disinterested men to examine the case, and simply revise, or confirm, in the matter of *damages*, and nothing else. We have found juries to be very unreliable in such cases. If they can agree, they are almost sure to decide unfairly in favor of the claimants for more damages.

It is also suggested that all the money, or means, raised for highways should be collected like other taxes in "*legal tender*," and none of it in labor. This "working out taxes," as sometimes allowed, is a challenge for shirk and laziness, and a very poor financial scheme. Let the work be advertised, and given to the lowest responsible bidder, so far as it can be done by contract; and put the day work under the charge of thorough and experienced foremen, and the competent superintendent will then make the most of the means at his command.

The importance of some such plan as we have suggested for the skilful engineering and superintendence of our public roads can hardly be exaggerated. The proof of bad management is apparent everywhere in our poorly constructed, and more sadly neglected, public ways.

We are confident that the plan would prove a financial economizer, rather than a bill of expense; and that the people could thus be gradually educated to appreciate better roads, and be made willing to undergo still heavier taxation for the perfection of the system.

It is not intended to suggest in detail the constitution of the "board of commissioners," or the "corps of engineers," above named, but simply that the State does not need any highly honorable and dignified commission to make lengthy and very scientific reports, or a corps of distinguished engineers to ride through the country and give orders, but only a few sensible and practical men, who will understand their work, and do it in the best and most economical manner.

Other Reasons for Highway Improvement.

With the great lines of inland traffic taken up by railways, no less important are the feeders of these, and the supporters of numerous smaller centres of trade, which our highways must ever constitute. Nothing more surely builds up a country town than good roads radiating into a productive farming district. The interest of the buyer and the producer are one in this matter.

Improved roads to market increase at once the value of farm produce, for larger loads can be carried in less time ; commercial fertilizers will cost less, as well as every article that is bought for home use. Hence the value of farming estate is enhanced, the more remote from market sharing more equally with the nearer, and the proprietors find themselves paid back fourfold for their road taxes.

A few years since there was a great demand for branch railways, and they were built to some extent, of the same general character as the main lines, and run by the same rolling stock. Of course they have not paid the railway companies, and the towns, or individuals, that helped build them, have not generally received back an equivalent in other profits. All such facilities, as a rule, increase the trade of large centres, and impoverish the smaller ones. The present tendency to great monopolies, and centralizations of trade, do not seem to be promotive of *general* prosperity, intelligence, happiness, or good morals.

Now every town of importance demands not a *branch*, but a *through* line, and they are obtaining authority to raise five per cent. on their valuation for this purpose. We fear many of these investments will prove *sinking* funds, without the hope of resurrection. But something must be done ; there must be approximate equalization of transportation facilities. If these cravings are morbid, they must be satisfied in some way, rather than denied.

Are we not overlooking the real cause of the difficulty? Supposing we had such roads as enlightened and practical science, with a wise and prudent expenditure, might have given us, would not the case be far different ?

But we are just at the dawn of a new era in railway construction and steam road-traffic. On our main lines, even the waste of carrying so much dead weight has for some time been discussed. But, without questioning the economy of such ponderous rolling-stock, and the need of hauling so many empty cars on our main routes, or doubting the propriety of hauling magnificent hotels, with kitchens, dining-halls, parlors and staterooms, at the rate of twenty-five or thirty miles an hour, from New York to San Francisco, we simply wish to notice a new system coming into use for branch lines and those of inferior traffic. With light steel rails, or wooden ones, if the others cannot be afforded ; with smaller and lighter cars and locomotives ; with some idea of making the paying load contribute to the adhesion of the engine, rather than so much dead weight of iron, by combining the power and vehicle in the same carriage ; with tracks laid at the centre or sides of our improved highways, or with broad driving-wheel tires running directly upon the road-surface, on grades as high as the maximum that should be allowed for first-class roads ; and in short, using steam on our public roads as familiarly, as safely, and with as little annoyance, as we now use horses.

Already steam omnibuses are doing good service in the streets of Paris, injuring the pavements less, and overcoming heavy grades more easily, than other vehicles ; and steam is hauling heavy loads with equal success, not only in London, but on highways in benighted India ; and nothing prevents its use here but our poor roads.

The speedy improvement of our public roads, then, not only for ordinary traffic, but to prepare the way for the introduction of steam on the more important lines, will best meet the wants of the towns not on the great thoroughfares, and enable them to put their money to better use than in stocks of expensive railways, which are sure never to pay them dividends, from the very fact that they are called upon to invest in them ; and there are *chances* that the big iron way may prove a way *out*, rather than *in*, for their business.

AMHERST, January 28th, 1870.

THIRD PRIZE ESSAY

ON THE

MAINTENANCE AND REPAIRS OF COMMON ROADS.

By Henry Onion, Civil Engineer.

Introductory.

In order rightly to perform a piece of work, or to accomplish a particular purpose, it is necessary, first, to have a clear understanding of the purpose intended, and then to consider the means of accomplishing it.

There is an urgent demand for improvement in the condition of the common roads throughout the country, and in this State, particularly, the demand is becoming imperative. To ascertain what is required to effect this improvement, it is necessary to look at the defects of the roads as they are now kept, and to mark the difference, in point of construction and maintenance, between good and bad ones in order to know what remedies to apply; also to examine the present system of management to discover where its faults lie, so as to provide understandingly some plan which will secure the most useful results with the best economy.

Present State of the Roads.

The public prints everywhere make frequent complaints of the roads at all seasons, and upon the breaking up of the winter and the roads together, the travelling is described as "horrible." All sorts of suggestions, many of them impracticable or worthless, and some spasmodic efforts, are made to remedy the evil, but generally with very little beneficial effect.

The roads are everywhere bad; difficult to haul over and unpleasant to travel upon. Their defects are obvious to every one.

Except in a few of the summer and autumn months, when they are more or less dusty or littered with loose stone and rubbish, they are muddy, deeply rutted, and frequently, after heavy rains, badly washed and gullied. Water often stands in puddles in the ruts and hollows, sometimes covering the whole surface for considerable distances. They are more or less rough, stony or sandy at all times, and when newly repaired, the material is so applied, or is of such quality, that for long periods travelling is more difficult and dangerous than before any attempt at improvement was made.

They are, moreover, often unnecessarily hilly or undulating, sometimes too wide for economy, but more frequently too narrow for convenience or safety; besides, the surfaces are generally so formed as to invite or compel the travel to follow a single track, confining the wear to one part, instead of distributing it over the whole breadth.

Importance of Good Roads.

We hardly need to urge the importance of good roads, for they are almost as necessary to the existence of a civilized community, as houses are for people to live in.

Without roads we should never have emerged from barbarism; and every improvement upon them contributes to the advancement of the people.

The common roads are the principal means of communication between neighbors, more or less near or remote, of facilitating an interchange of good offices and new ideas; they also permit of an exchange of commodities, and thus in all ways promote the intelligence and prosperity of communities.

They are, besides, tributary to the railroads, carrying the surplus products of industry to the nearest stations, to be transported to distant markets, and, in turn, distributing to us at our doors the freights brought back for home use.

The area benefited by railroads and local markets is extended within certain limits, in proportion to the goodness of the common roads. A man living four or five miles from a market or a railroad station on a very good road is practically nearer than one living two or three miles distant on a bad one.

At the West the writer has often seen two or three yokes of oxen employed in hauling loads that would, on a good road, have been light burdens for single horses.

Characteristics of a Good Road.

The characteristics of a good road are hardness and eveness of surface, with a degree of smoothness which will enable carriages to move with the greatest ease, while affording a sure foothold for horses; the hardness being sufficient to resist the pressure from loaded wheels and the hoofs of animals going over it, but not so great as to prevent entirely an elastic yielding under moving loads. The art of road-making consists, in a great measure, in applying the best means to secure these qualities, and in such a manner as to insure durability against the wear of use and the action of the elements.

Drainage.

The requisite foremost in importance, attention to which cannot be too urgently insisted upon, in the construction or improvement of roads, is thorough drainage; for without this a durable or even a temporarily good structure is hardly possible. Neglect in this particular is one of the main faults of our method of treating roads.

Provision must be made not only for the removal of surface water, but of all excess of moisture from the substratum below the roadway. Under-drains should be laid wherever the soil is not sufficiently porous for self-drainage, and the water tables and side ditches should be so arranged as to carry all surface water quickly away.

The under-drains should be put down in the same manner as for agricultural purposes—say about three feet below the surface, and from eighteen to not more than forty feet apart, according to the character of the soil and amount of water retained by it. Generally, these drains would be arranged longitudinally with the road, but in some cases they may be put across it, in the form of a flat V, the angular point highest, or up the incline on descending grades. This work must be executed with judgment and careful attention to details, to be effective and to save unnecessary expense.

The side channels, or water tables, for receiving the water from the road-surface, should have a fall of as much as 1 in 30; but the inclination should not be so great as to produce violent currents, and at short distances means should be provided for discharging them of water, to prevent an overflow, or too great a current to the hazard of the road.

Side ditches should be made wherever required, with sufficient capacity and fall to carry away quickly all surface water from the near vicinity of the road, and should have a depth of three feet below the surface.

Catch-water drains are sometimes necessary to prevent encroachments of water from neighboring hill-slopes and the sides of excavations. Suitable out-falls for the drains of every description must be provided, or they may become useless, or perhaps destructive of the works they are intended to preserve.

Culverts.

Culverts for carrying streams under the roadway should be of sufficient size, and be set deeply enough not to impede the natural flow of the water, and, if needed, to help drain the soil near and under the road. It would be better always to make them large enough for a person to enter in a stooping posture, so as to clear them of chance obstructions. Small culverts are liable to become choked, when it often becomes necessary to remove their coverings to be able to clear them. A very common fault is to build them in such a way as to form little ponds on the up-stream side of the road, thus often keeping the embankment saturated with water for long periods.

Effects of too much Moisture on Roads.

It may be thought that so much care about drainage is unnecessary ; but when we consider that the effect of too much moisture is to soften and loosen the soils and all kinds of material used for road-covering, it will be perceived that to this cause is mainly attributable the bad character of our roads. The difference in their condition between spring and summer is an evidence of this fact.

If kept dry at all seasons, the wear of the surface would be very much diminished, and the cost of repairing material consequently reduced.

When the covering of the road is compact and firm, so that the fragments of which it is composed are held fast in place, the wear is necessarily all on the surface, and, if made of good material, it is very slow ; but let it become loosened and saturated with water, the fragments will be displaced by the wheels of carriages and the hoofs of animals, which will crush and grind them together, so that they soon become rounded or reduced to powder, and need to be replaced by new material.

If roads are thoroughly drained, the action of frost upon them would be very much diminished, and the annual breaking up of their surfaces, produced by spring thaws, would be prevented. The effect of heavy rains, frequently so destructive to water-soaked roads, would be the most efficiently guarded against by means of thorough draining.

Trees and Shrubs on the Roadside.

The roads should have a free exposure to the action of the sun and the winds, in order that moisture may dry off quickly. Hence, natural hedges and trees ought not to be allowed to line their sides in a way to interfere with this object. The dripping from branches, too, hanging over the road is injurious.

Besides overshadowing and sheltering the road, shrubs and trees are often allowed to encroach in such a way as to discommode the travel.

Width of Way.

The width of way between the larger towns, or where the travel is considerable, should be thirty feet, increased, perhaps, to forty feet at a near approach to the towns, but ought never to be less than twenty-five feet.

The first cost is somewhat greater for a wide track; but if the width is not extreme, and the surface is properly covered and shaped, the same amount of material will wear much longer than upon a narrow one.

Upon narrow roads the travel is apt to be confined to one part, thus producing ruts and unevenness, which hastens the wear, while upon a wider one this is not so likely to occur.

Cross-Section of Roads.

The form of cross-section is very important, and one upon which there has been much difference of opinion, some advocating a flat surface, others holding that the surface should be very crowning. These latter urge the necessity of making the centre of the road much higher than the sides, because the travel goes there, and consequently the wear is mostly in that part. But it is because the surface is crowning that the travel does go upon the centre, for the carriages can stand upright in no other position.

The only useful purpose served by raising the centre is to allow the water to run off at the sides.

If, however, too much inclination is given to the sides, the travel will seek the centre, as before remarked, and that part will soon be worn into ruts and hollows, making conduits and basins for the water there.

For these reasons, practical engineers generally coincide in the opinion that the centre should be but little higher than the sides, and that the best form of cross-section is formed by two straight lines inclined to the sides, connected at the vertex by an arc of a circle of about ninety feet radius. The inclination of the sides ought not to exceed 1 in 30.

The common practice with us is to give the surface a cylindrical form, making the cross-section nearly the arc of a circle, with a short radius, so that the inclination increases from the centre towards the sides. This is the worst possible form, especially when the road is narrow, or, as is usual, the centre is raised very high; for, in order that vehicles may retain anything near an upright position, they must go upon or near the middle of the way, confining the wear to that part; but if the amount of travel is sufficiently great to compel carriages frequently to turn, or to keep upon the sides, they are exposed to accidents by upsetting, and the tendency to slide down the slope increases the wear of the wheels and of the road.

Besides, the labor of the horses is increased by this tendency of the wheels of carriages to slide in a direction at right angles to the line of draught, and the chances of breakage are multiplied by the augmented strain upon their axles.

Gradients.

The longitudinal inclination of roads, where necessary, should be as slight as possible, and ought never to exceed the angle of resistance for the materials of which the surface is composed.

The results of numerous experiments to determine this angle, or the angle of inclination, for different materials, which is just sufficient to cause a carriage standing upon the inclination to commence moving down it on the slightest application of force in that direction,—are repeated in nearly all treatises upon road making, and need not be re-stated here. The useful deduction from them and from practical experience, however, is that for a well-made road, with a hard and compact surface, the inclination ought not to exceed 1 in 30, and for ordinary gravel roads 1 in 20.

These inclinations are sufficiently safe, but a vertical rise is equivalent to an increased length of road proportional to the angle of inclination. From calculations made by means of a formula deduced from experiments tested by Sir John Macneill, it is shown that a goods wagon of six tons burden, drawn three miles an hour upon an inclination of 1 in 30, one mile is equivalent to 2.7 miles level road; and for a stagė-coach of three tons drawn six miles the equivalent level road for one mile is 1.62 miles.

The amount of force expended in conveying a given load over a road from one point to another at a higher elevation, is equal to the force of traction plus the force necessary to lift it up to the elevation reached: thus if the load is carried two miles, and the terminus is a hundred feet higher than the starting point, it has to be *lifted up* that hundred feet, as well as hauled the two miles.

Any intermediate descents will not compensate for the rise, but will add, by carrying the load to a lower level, to the ascent to be made.

The difficulty of making the ascent will be in proportion to its inclination, and it will be necessary to start with a team of sufficient strength to climb the hill, which would be greater than would be needed on a level, or descending road.

The road-covering upon steep inclinations is liable to be torn up by the hoofs of the animals making extra exertions to ascend with heavy loads, and is more subject to injury from deluging rains, in proportion to the pitch of inclination. Hence hills should be avoided or reduced as much as possible.

These considerations are apt to be overlooked by persons ordinarily having charge of improvement, or the construction of roads, or otherwise they might be amended by material reduction of grades, and generally with very little extra cost.

12

Grading of Roads.

The finished road consists of two distinct parts: the sub-road or road proper, and the covering. The sub-road is first graded up to within about a foot of the intended finished surface, and its surface shaped and prepared to receive the covering.

It is also important that the embankments should be made in a solid and substantial manner, care being taken not to put into them vegetable matter, or decaying substances of any kind to endanger their stability, or to cause unequal settling.

In cases where the sub-grade runs near the surface of the ground, the sods must be taken off, and all roots and stumps removed.

Mucky earths must be replaced by material fit for a solid foundation.

Ledges should be excavated to a depth of one foot, at least, below grade, and the surface dressed so as to leave no hollows for holding water.

The sub-road should have the same cross-sections as the surface of the finished way, so as to shed any water that may percolate through the superficial coating, and in order that the covering may have the same thickness throughout.

Foundations for Road Covering.

That part of the structure lying immediately under the road covering may be termed the foundation, and upon it, in a measure depends the durability of the covering, and the ease of travel.

As shown by Sir John Macneill's experiments, before mentioned, the force required to move a ton, on a broken stone surface, on a bottom of rough pavement, is forty-six pounds; while upon a broken stone surface, laid on an old flint-road, it is sixty-five pounds.

The covering may be laid directly upon the ground without further preparation, than above indicated, or a foundation may be made of concrete, or of rough-paving.

Many of the military roads of the Romans were constructed upon a concrete foundation, some of which have endured to this day without being entirely worn away.

The use of this foundation, was introduced into England not many years ago by Thomas Hughes.

It is made of gravel and lime in the proportion of five or six parts of clean gravel to one of lime. The lime is finely ground, the materials thoroughly mixed, and the concrete is made on the surface of the road.

The depth of the concrete bed when applied is six inches; upon which a layer of stone or gravel three inches in depth is spread, before the concrete has set.

Afterward, and before the travel is permitted to go over it, another layer of material is spread, and the whole consolidated by rollers.

In some instances the concrete foundation has been successfully tried in cases where no other means were effectual in making the road solid.

The method of making a foundation of rough paving introduced into England by Mr. Telford has been extensively used in Europe, and to some extent in this country.

This is formed by laying down broken stone of not more than twelve inches in their greatest dimensions, nearly parallel-sided, and not more than seven or eight inches in depth.

The stones are laid close together broadest side down with the longest way across the road, and then are wedged together with thinner stones and chips, after which projecting points are broken off with a hammer, and the surface evened by filling the hollows, and crevices with stone-chips.

Road-coverings based upon foundations of paving, or concrete are so superior, and so much more durable as to justify their use wherever the traffic is large, and the saving in the wear of material and the labor of repairs will, in time, more than compensate for the extra cost of constructing.

They receive the pressure from loads, transferred by the fragments composing the covering, without yielding; thus checking any tendency to movement amongst them, and prevent their being forced into the soil below, and the soil from working up, and mixing with them.

With such foundations a less amount of material will be required for the covering, which need not exceed five or six inches in depth, whether made of broken stone or gravel.

The covering placed upon the earthy surface, with no intervening stratum, will need a depth of eight or ten inches, and cannot be worn so thin as when upon an artificial foundation without re-coating.

The purpose of covering roads is to obtain by the use of appropriate materials a uniformly hard, even, and unyielding bed, that will resist wear, and afford a smooth surface, which will offer the least resistance to the motion of the wheels of vehicles.

Covering of Roads.

As the covering receives the shock, and suffers the attrition from the hoofs of animals and the wheels of carriages, it needs to be made of such substances as will best resist crushing and abrasion.

The excellence of the road depends upon the characteristics of the covering.

The materials usually employed in forming it, except in large cities, or where the traffic is very heavy, are broken stone and gravel.

The foundations of concrete and stone, when completed, are ready to receive the coating ; but it would be a great deal better to prepare the earth foundations by compressing the surface as much as possible with heavy rollers.

The stone used for covering should be selected with care, to secure such as is both hard and tough, and broken into angular pieces, such as may be passed through a ring two and one-half inches in diameter and down to one-third of that size, but no smaller.

The broken stone should be kept clean, and in handling, a pronged shovel should be used, to prevent dirt from getting mixed with it.

When used on artificial foundations, it is to be spread evenly in layers over the whole breadth of the road, the bottom layer three and one-half inches in thickness.

The first layer ought then to be compressed firmly with rollers sufficiently heavy for two horses to drag, when the top coating, of two or two and one-half inches, should be applied, and rolled down in the same manner as the first.

Finally, a coating, of about an inch in thickness, of small gravel should be spread over the surface, and the whole consolidated with rollers weighing from six to ten tons.

During the rolling, the surface must be kept even by filling any hollows or raking down any bunches that may make their appearance.

The object of covering the surface with gravel is to fill the crevices between the fragments of stone, to help hold them fast in place, and to reduce the shock upon the wheels from striking the angular points of the stones, by preventing them from falling into the cavities. When finished, the surface should have solidity and smoothness sufficient for a horse and carriage to trot over it with ease.

The same process is required for covering a road-bed which has no artificial foundation upon it, except that the *thickness* of the layers should be increased, or the *number* greater. Gravel consists of small fragments of stone, more or less rounded, and as found inland, is generally mixed with sand or loamy earth, while that obtained from sea-beaches is usually free from extraneous matter. It varies greatly in quality, and it is difficult to find, in its natural state, suitable for road-covering. For this use it should contain no pebbles of more than one and one-half inches, or less than three-quarters of an inch in diameter. It should be free from sand, but must contain a proportion of binding material of a loamy or clayey character.

It is generally necessary to prepare it for use by screening, to separate from it both the larger pebbles and some of the finer stuff, when that is in excess. When gravel is of excellent quality, or it is properly prepared, it makes a superior road-covering, for which purpose it should be spread in layers and consolidated in the same way as that described for broken stone.

The covering materials should be kept moistened, by sprinkling or by the state of the weather, while being compressed under the rollers, but the rolling should not be done when they are saturated with water.

It is usual to admit the travel upon the loosely spread materials, and make it do the work of binding them into place, which it will never do so perfectly as might be done by rolling. The wheels act as wedges, forcing themselves between and displacing the fragments which are crushed and abraded under the loads; mud and dust works into the interstices, so as to prevent the parts from wedging or being bound together; horses' feet are liable to injury from loose stones; the rims of wheels are rapidly worn; riding in carriages, too, is very uncomfortable, and without extreme care, the surface will become rutted and uneven. Thus, before the road has become settled, the discomforts and disadvantages outbalance the cost of rolling.

The use of the steam stone-crusher and the introduction of the steam roller would greatly reduce the cost of preparing and applying road materials; and the employment of these and other machines of like character would be a step towards great improvements in the common roads.

Relative Merits of the different Road Coverings.

Having briefly indicated the different methods of construction, it is proper, before considering the subject of repairs, to compare the relative merits of each.

There are three points especially worthy of consideration in determining what kind of road to build; or rather, as that is the important feature, what kind of covering to use, namely: the first cost, the cost of maintenance, and adaptedness to the use intended.

The fitness of a road for its use of course depends upon the kind and amount of traffic going over it; for one much travelled by heavily loaded teams will need a more substantial and expensive covering than one used chiefly for light carriages; and consequently judgment is required to avoid deficiency of strength and solidity, on one hand, and unnecessary expense on the other.

The comparative original cost of covering a road would depend in some measure upon the kinds of material most procurable in the vicinity, or the distance to where the kind needed must be sought; but generally the broken stone covering, with a paved or concrete foundation, is the most expensive. The next in order, as regards the first cost, is a broken stone covering laid upon earth, though the difference of this from the first mentioned, in that respect, is not great.

Gravel is the cheapest of all materials for finishing roads, when it can be obtained, of good quality, from a reasonable distance, and can generally be most easily applied.

It would appear, however, from experiments made by Mr. Wm. H. Grant, superintending engineer, upon roads constructed by him in Central Park, in New York, that the relative cost of them was for

Gravel upon a paved foundation, ...	1
Broken stone upon a paved foundation, ...	1.65
Broken stone upon earth, ...	1.70

Broken stone roads are designated, respectively, those having paved foundations, Telford, and those laid upon earth McAdam roads.

For general traffic, the Telford road is undoubtedly the most durable, and can be maintained in high condition, at less cost, than any other. The foundation is permanent, and, with a proper depth of materials perfectly consolidated, the whole will always remain intact, except from the wear upon the surface.

The macadam way is more liable to be broken up from the yielding of the earth foundation on which the materials rest, thus subjecting them to a tendency to motion among themselves under heavy pressures. Any uneasiness of the fragments causes the soil to work upward, so as to help loosen their bonds ; therefore the wear would not be confined to the surface, and the necessity of replacing them would be more frequent.

Gravel upon a paved foundation, thoroughly consolidated before use, doubtless makes the best and least expensive of really good roads for light carriages. They are well adapted to localities where there is but little heavy traffic, and can probably be kept in repair, in such situations, with as little cost as any other.

According to a statement made in the "Engineering and Mining Journal," by Mr. Grant, before quoted, a sample road made in Central Park in this way, remained in perfectly good condition, after nearly five years constant use, during all of which time it has required no repairs, except at a point where the gravel became loosened and uneven by the turning of carriages about a short curve.

Roads made of gravel on earth foundations are less fitted for general travel than any of those mentioned, and though cheaply constructed are the least useful and most expensive in the long run, requiring a frequent renewal, or constant patching of the covering, to keep them in tolerable order.

Earth roads made and repaired with plough and scraper are hardly worthy of mention except as temporary expedients in new settlements, though they are not uncommon in the oldest parts of the country.

In economy of repairs, the roads having artificial foundations for the coverings, and otherwise thoroughly constructed, are superior to any others. The saving in the expense of their maintenance will generally in a little time compensate for the excess of original cost. In economy of use they are superior in offering less resistance to draught, thus allowing greater speed to be made and heavier loads to be carried with the same animal power.

Methods have been contrived for measuring with accuracy the force required to haul loads over different kinds of road surface, or up and down different inclinations at the various ordinary speeds of travel for all sorts of vehicles used. With these many experiments have been made, and the results generalized by means of formulas so as to be applicable to any particular case.

Some of the results obtained from experiments made by M. Morin at the expense of the French Government, are as follows:—

1. The traction is directly proportional to the load, and inversely proportional to the diameter of the wheel.

2. The width of the tire, if above three inches, does not affect the traction upon paved or hard macadamized roads.

3. There is no difference of traction for carriages with or without springs going at a walking pace on the same road.

4. Upon hard macadamized, and upon paved roads, the traction increases with the velocity; the increase of traction being directly proportional to the increase of velocity if greater than two and one-fourth miles per hour. The increase of traction arising from an increase of speed is less in proportion to the smoothness of the road, and the lightness with which the carriage is hung.

5. Upon soft roads of earth, or sand, or turf, or roads fresh and thickly gravelled, the traction is independent of the velocity.

6. The traction, at a walking pace upon a well made and compact pavement of hewn stones is not more than three-fourths that on the best macadamized roads.

7. Carriages without springs are more destructive to roads than those with them, and the less the diameter of the wheels the more destructive they are.

Sir John Macneill, with an instrument for measuring forces required to haul loads over different surfaces, obtained, from many experiments, the results contained in the following table; inserted here for purposes of comparison. The wagon used weighed twenty-one cwt., and the resistance to draught was:—

On a well-made pavement, ...	33 lbs.
On broken stone laid on pavement or concrete foundation, ...	46 "
On broken stone laid on earth, ...	65 "
On a thick coating of gravel laid on earth, . .	147 "

These and many other like experiments made in France and Great Britain point clearly to the great superiority of the best broken stone roads, over those made of gravel and loam in the ordinary way. More will be said bearing upon this subject further on.

Suitable Persons to have charge of Roads.

To secure such kinds of structures as have been described, or to make material improvements upon those in existence, it ought not to be necessary to intimate that the work must be done under the direction of practical road-makers, whose education and training fit them for such duties. But the common practice of entrusting such work to those who are not only wholly ignorant of the mechanical principles necessary for the achievement of success, but who have had no experience useful for the duties required, seems to indicate that this consideration is not appreciated.

An extensive knowledge of scientific principles is necessary for the successful practice of the engineering art, of which road making is not the least difficult or important branch. An acquaintance too, with the processes and expedients, whether successful or not, employed by men who have achieved eminence in their profession, is needful to help mark out the right path, and to hint at the direction in which to look for resources in difficult emergencies. No art can, however, be learned from books alone, but the skill to execute, the ability to judge of the qualities and fitness of materials, the ingenuity to overcome difficulities, and the readiness to find and adopt methods of operation suited to each particular understanding must be acquired from practice.

A cavalry man detects at a glance the good points of a horse;

no defects are concealed from his scrutiny; he knows how to correct his faults, how to feed, and groom him, so as to fit him for service, and he can guide and manage him in a manner to make him most useful. All this is a mystery to the foot-soldier who can see none of the beauties, or faults of the horse; he knows nothing of his qualities, and can hardly distinguish one animal from another.

None but a farmer will undertake to breed stock, or grow crops; a blacksmith does not attempt to make glass ware; then why should an ignorant day laborer, as often happens, be employed to make and repair roads?

Improvement and Repair of Roads.

It has not been thought necessary to state in detail the process of doing the work, or to describe minutely the materials to be used, to do which would require a volume, as there are many excellent treatises upon road-making, which are easily accessible to those who wish to pursue the subject. The object here is, as clearly as possible, in a brief space and limited time, to point out the public needs in regard to roads, and to suggest the best means of supplying them.

In pursuance of this purpose, we will now, at the risk of some repetition, proceed to state, in a general way, what is required for the improvement of the present roads, and to consider what ought to be done to maintain them in first-rate working condition.

Many of the old roads upon which reduction in grades, or changes in direction are not needed, may be easily improved. With some trimming of the sides, shaping, and evening of the surface, being solidly compacted by years of travel, the beds of these roads will serve as a good foundation for covering with either of the kinds of materials mentioned. If gravel is used it must be carefully selected, and a coating of the material as it comes from the pit, only removing the larger stones, should be spread evenly to the depth of three or four inches over the whole surface. It should then be raked smoothly, and solidly compressed by a roller drawn by two horses.

Over the first layer after it has been sufficiently hardened, a second one two and one-half inches in depth, consisting of screened gravel, is to be put on and treated in the same manner, when the whole should be thoroughly consolidated with rollers of five or six tons weight. If then the drainage is properly attended to, the roads will become nearly equal to the best, and with watchful care, may be kept permanently so, with small cost, in comparison to the present outlays for patching and repairs. A saving might be made by putting on both layers unscreened, but the road would not be so perfect or durable; besides this, the greater ease and comfort of travel upon the screened gravel, and the difference of wear would doubtless repay the cost of screening.

Upon places where the old road is not sufficiently firm—being either sandy or loamy—a foundation of rubble-stone may be laid to the depth of eight or nine inches. The stones used for this purpose should be angular, having no dimensions greater than twelve inches or less than three inches, and should be carefully and closely laid together, the longest way across the road.

The surface should then be evened by spreading stone chips over it, when the gravel may be put on and treated as described above, except that an additional layer will be needed to guard against a movement among the loosely deposited stones below.

When the roads are already in tolerable good condition, with a good gravel surface, nothing more perhaps will need be done than to dress them into proper shape and to fill the hollows and even the surface. If however the surface is much worn into ruts and hollows, or the gravelly coating has become thin, a layer of two or three inches of screened gravel should be spread evenly over the whole of it and rolled solid.

In situations where gravel of suitable quality cannot be obtained, or upon roads much frequented by heavily loaded teams, broken stone must take its place in the work of renovation. In such cases the surface of the road must be newly shaped, in the same way as for gravel, taking care not to break it to a greater depth than just sufficient to give it the proper form and to make it even. The manner of applying this material to the old road-beds is the same as has already been described.

In all cases, the mode of proceeding and the precautions to be taken are the same for the old roads as for the new. Wherever the travelled part of the road is not of sufficient width, it should be made so before being coated with surface-materials, and the sides or centre, as the case may require, should be reduced to obtain the proper cross-section.

Shrubs and trees growing upon the borders, not intended for ornament, should be uprooted, and the heaps of stones and rubbish that so commonly offend the sight and sometimes obstruct the way, should be removed.

The Cost of Improvements and their Benefits.

No one acquainted with the present wasteful and thriftless way of dealing with the roads, can doubt that they might, under a more effectual system of management, be gradually brought to a high degree of excellence, without an increased outlay of expense; and some facts will be presented hereafter, going to show that they can be maintained, in a superior condition, at even less than the present cost.

The importance of bringing up the public roads to a high standard of excellence, and of constantly keeping them in that condition, has been sufficiently urged, and is perhaps generally well enough appreciated; but the possibility of doing so with reasonable economy, or without, as might be apprehended, a largely increased expenditure, might, on the first thought, be questionable.

But we have seen how a great saving from the present expenditure of money and labor upon the roads might be made by a proper application of the work, and that any additional cost contracted for construction would be, as well for their better adaptation for use as for the purpose of making them, very much more durable, and consequently as much less expensive to keep in repair.

Besides, the well-made and well-kept roads would be greatly superior for the purposes of transportation and travel, since very much heavier loads could be conveyed by the same expenditure of animal power, and valuable time be saved by higher speed.

It would be safe to say that a saving of twenty-five per cent. in animal power alone would result from the proposed improvement of the roads. In these considerations, the gain from diminished wear of vehicles and harnesses, and the incalculable saving of oaths, impatience and ill-temper, have not been taken into account.

There can hardly be a doubt, from these considerations, that the people could afford to pay even an increased tax for extensive improvements and the support of all the public roads, and still be gainers by the reduction in the cost of traffic that would result.

Roads in Europe.

The advantages of improvements in public roads by skilful reconstruction, and by watchful and unintermittent care for their preservation, has been tested in several countries; and on the Continent of Europe systems of intelligent superintendence have long been established, always with the best economical results.

The following statements of facts are quoted by Gen. Sir John F. Burgoyne, in his "Remarks on the Maintenance of Macadamized Roads," from papers written by Monsieur L. Dumas and other French engineers of the Corps des Ponts et Chaussées:—

"The following took place with respect to the high roads (Routes Royales,) of the Department de La Sarthe, somewhat less than two hundred and fifty miles in extent. In 1793, a demand was made to put them in complete order, £15,280, or £60 per mile; in 1824, the demand was £9,000, or £36 per mile; in 1836, the demand was £7,760, or £31 per mile; in 1839, the demand was £6,640, or £26 per mile; and the roads have become better concurrently with the reduction of cost of maintenance, from being, in 1793, in deep ruts, to 1839, when they were in very good order.

"Part of the great road between Lyons and Toulouse, till 1833, was always in a dreadful state, and yet cost habitually about £110 per annum per English mile for maintenance, when M. Berthault Ducreux introduced a system of patching, instead of general repairs, since when the road was greatly improved, till it was in a *very good state*, and the annual expense reduced by £13 or £14 per mile.

"Another instance is quoted, where, prior to 1837, the average amount of broken stone laid on thirty-three miles of road was 6,000 cubic yards; under the improved system in 1837, 5,000 cubic yards were applied, in 1838 only 1,350 cubic yards, and 1839 *none*. In another instance the cost has been as follows for the expenditure:—

YEARS.	Materials.	Road Labor.	Total.
1830, ...	£548	£166	£714
1831, ...	563	188	751
1832, ...	496	151	647
1833, ...	500	167	667
1834, ...	438	177	615
1835, ...	398	145	543
1836, ...	380	160	540
1837, ...	360	178	538
1838, ...	180	233	413
1839, ...	250	300	550

"In 1837, when it was taken up in the new system, the road required considerable improvements; in 1840 two-thirds of it were in perfect condition, and when the whole is reformed, it is calculated that from £400 to £440 will keep it perfect.

"The road from Tours to Caen in La Sarthe was, in 1836, in so bad a state as to be in danger of becoming impassable.

"In January, 1837, it was passed under the charge of M. Dumas. The average annual expenditure for the five years prior had been: for material, £736; for road labor, £242; total, £978. From 1837 to 1841 the annual average for material was £375, labor, £443; total, £818. In 1841 the cost of material was £163, for road labor, £445; total, £608. In 1838 this road was reported to be in a very good state, and since then has become better and better.

"In 1834 the mail required always *five* horses, and the road was so bad that the postmaster *lost eleven* horses by the hard work in *one year*.

"In 1838 the number of horses was reduced to *three*, and at present there are only *two* of *middling* quality, and the postmaster loses *none* from that cause.

"In this same district, in consequence of improvements in these roads since 1839, a number of lighter carriages have been established; they have four wheels and are drawn by one horse, carry nine persons, and go seven or eight miles an hour. Previously the carriage for the same number of persons, had two wheels, two horses, and went much slower."

At the time these statements were made, these French engineers calculated that by maintaining the roads in the best possible condition, which they assert can be done without increased expense, the cost of draft of merchandise over the roads in France might be reduced one-third, or about $30,000,000, saving that amount to the public. From the foregoing statements it will be perceived that the cost of labor upon the roads is increased, while the cost of material is diminished by the system of constant care.

The inferences to be drawn from them, and from experimental tests, all point the same way as the suggestions of reason, and from experience gathered from the practices observed in this country, namely, that the best public economy and convenience would require that the highways should be maintained in the best possible condition. The cost of maintenance including improvements, would not be increased, the wear and tear of carriages and harness would be diminished, there would be a saving of time, by increased speed, less cruelty to animals, and far greater comfort in travelling.

Defective System of Management.

It is useless to attempt to improve the condition of the roads except with a change of system, or rather by the substitution of a system for the hap-hazard way the work is done. Constant care and oversight is required, and workmen should be employed to be on the road all the time, to keep the surface clean, and to repair any injury or defect as soon as it occurs.

The importance of cleanliness is illustrated by some experiments made by M. Morin, showing that the resistance to draught upon broken stone roads is four times as great with deep ruts and thick mud, as upon a good one, and is greatly increased by the muddy or dusty condition of otherwise perfectly good roads.

The way Roads are now Mended.

The laws of Massachusetts relating to the maintenance of the highways, however well adapted to *colonial times* when the country was thinly settled, and money hardly to be had, are totally unsuited to the present age with our dense population and abundant means. The towns are required to choose annually one or more surveyors to make repairs on the roads, and the custom is to elect from one to twenty or more, according to the territorial extent of the town, or the caprice of the voters.

To each surveyor is assigned a certain portion of the road-tax either in money, or in days' labor, according as the town determines that the tax shall be paid, and he proceeds to expend the same upon the roads allotted to him, exercising his own discretion about the time and manner of doing his work, and using his own judgment, if he has any, of the kind of work required.

The time he chooses is generally when he has the most leisure, and whether required or not, he frequently works out his money at once, and gets done with it. The work is often performed in the rudest manner, and the road is coated here and there with thick patches of worthless stuff—better suited for top-dressing for crops, perhaps, than for road material—to be washed into the gutters on the occurrence of the first heavy rain. From want of judgment, or want of interest, the money is wasted, and the people are burdened with the heavier tax of struggling over hard roads, made worse by the money they have paid for improving them. Less frequently a man is chosen, who understands better what is needful to be done, and with what means he has, he commences some improvements, which if followed up in after years would result in a public benefit, but his office is only for a year, and he may be followed by a man to undo, what was well begun.

Our people do not manage their *private affairs* in this thriftless way, neither is any other public business conducted so loosely or wastefully, and it is a wonder that these evils have been allowed to continue so long as they have.

A few towns have tried to remedy these evils, with some measure of success, by appointing a superintendent to have charge of all the roads in the town, and keeping a corps of workmen, with horses and carts constantly employed in making repairs.

The public moneys are contributions from the surplus earnings of the industrious brains and hands of the Commonwealth, and should be expended in the best discoverable manner, to secure the greatest public benefit. The people ought to require that these important trusts shall be confided to none but such as are fitted by skill, intelligence, and integrity, for the duties assigned them; and the laws should be so framed that the money provided for important public purposes shall not find its way into wasteful and incompetent hands.

Proposed System.

For the efficient and economical maintenance of the public roads, it is essential that there be a uniform system of management common to the whole State. The first step towards a complete reform of system would be the creation of a State department of roads and bridges, to have general charge of all the roads, to arrange and direct the carrying out of the details, and generally to look to the effective working of the system.

The chief of the department should be a practical civil engineer, thoroughly conversant with the art of road making. For the purposes of proper supervision, the State might be divided into districts, say by counties, and these again into sub-districts, larger or smaller as might be found expedient.

There should be a resident engineer or superintendent for each district, to have charge and oversight of the roads and bridges, within his district, and to be held accountable to the chief of the department.

He will ascertain the condition of the roads in his district, determine what improvements are to be made and in what order, decide upon the kinds and amount of work to be done, estimate the sums needed to carry it on, and at stated periods report the same, with all other matters pertaining to his office, to the chief of the department.

For each sub-district there will be required an assistant-engineer or road-master, subordinate to the resident of the district, to manage the working details, within the limits assigned. As the improvements progress, these sub-districts may be enlarged and the number of subordinates reduced, so that each and all shall always have work enough to keep them occupied.

By some method of this kind the standard of excellence and the manner of treating the roads would be uniform; any suggestions of experience could be taken advantage of; the use of improvements in implements and machinery could be everywhere extended, and a system of strict accountability maintained throughout the department.

The public could thus enjoy the benefit of the skill and ingenuity of educated talent, which under town management is scarcely possible, and the people would be assured that their money would bring them beneficial returns instead of being wasted by mischievous incompetence.

Under a general system, stone-crushers, steam-rollers and other costly machinery might be employed, and other means of economizing labor adopted, that towns cannot command. Suitable materials for road making could be prepared at the most convenient places, or where found of a superior quality, without reference to town lines. Towns cannot individually provide for a competent supervision of the roads, an advantage which a general system insures.

The details of the system might be arranged as found most expedient, or as would best accord with the prejudices and habits of the people.

The tax levied for the roads might be paid into the State treasury, and the money expended for the best advantages of the general public, or allotted to the districts according to their relative needs, or it might be paid to the county treasurer, to be expended within the county, or again it might be collected as now, and expended wholly in the towns where raised.

Details of Working.

Either of two methods might be adopted for doing the work upon the roads. Gangs of hands working together may be assigned to districts of such extent as they can go entirely over in limited periods and do the work needed ; or men may be stationed singly along the line, each with the length of road that he can properly look after.

The workmen, in either case, would be subject to the oversight and direction of the road-master or overseer of the subdistrict. Parties working in gangs would each require a horse and cart for their use, and would do all the necessary work of cleaning and repairing the roads ; but the duties of preparing and distributing materials to depots convenient for use, of removing the road-covering where worn out, of rolling the surface, &c., would require to be performed by distinct parties. The materials to be used for patching, after being carefully prepared, should be hauled to points convenient for the use of the repairers, and deposited in depots prepared in such a way as to prevent earth or other deleterious substances from becoming mixed with it. These depots might be formed by enclosing suitable spaces by a wall of rough stone.

The men stationed singly,—the *French cantonier*,—under the system of assigning to every repairer a certain length of line, will each need to be supplied with a wheelbarrow, a pick, a shovel and a scraper.

These suggestions of details are only made to indicate how the proposed system may be made to work practically; but the experience that would be acquired under the system would in a little time lead to the simplest and most effective plan of operation.

How the System will Operate.

This system of management of the common roads which it is proposed to substitute for the one now in operation, is doubtless the best that can be devised to effect a permanent improvement of the highest character upon the roads, with the most economical expenditure of labor and money.

It provides for the responsibility of every individual having supervisory charge of the roads, by holding each in order accountable to one next higher in rank, and with a provision for a method of making payments at stated periods, upon pay-rolls and written contracts, by officers assigned to such duty, which should be included. Peculation and embezzlement would be guarded against, and a draining leak of some magnitude in the public purse would be stopped.

Under such a system, every officer of the department would, from a care for his own official and professional reputation, require that his subordinates should all be men of capacity and intelligence, who, from fear of dismission, if otherwise disposed, would not dare to be remiss in duties. If, however, there should be any hesitation about making all at once a change so sweeping, and the adoption of a new system should be delayed from any cause, it seems very desirable that a department or a commission should be established to act as an advisory board, helping the towns, if possible, out of some of their perplexities in regard to the care of their roads; to collect statistics and arrange them, and to gather facts and compare them, so as to furnish information, in a systematic form, upon which to base future legislation, and in such other ways, as may be found useful, to promote the object of improving the character of the roads throughout the State.

Reliable information upon nearly every particular in regard to the roads is needed. It would be useful to know the number of miles of roads there are in each town in the State, the cost per mile for keeping them in repairs, and how made, what materials are used, the condition of the roads respectively, &c.

It would be well, too, to know, in this connection, what, in the several parts of the State, is considered a load for a horse, and what is the average rate of speed, with fair driving, for light carriages.

There occur other useful inquiries which might be made, such as: what number of surveyors are usually chosen? whether steam-crushers or rollers are used? what damages, if any, the town has suffered from defects in roads and bridges? &c.

Intelligent answers to these and like queries, which the commission in question might be authorized to obtain, would naturally help the judgment in deciding what legislation is necessary.

Good Workmen and Neatness.

A reason not before mentioned why the system here advocated should be substituted for the present one, is that under it the laborers would be selected for their intelligence or aptitude, and those proving superior workmen would be retained permanently in the service; they would soon become expert in their duties, and do their work with celerity and neatness. The performance of a good workman will have a neat and finished appearance; or as the phrase is, a workmanlike look.

There is no excuse but stolid indifference, for the slovenly, and unsightly appearance of the borders of many of our roads. All rubbish, dead leaves, and loose stones should be removed from thence, and not allowed sensibly to accumulate. This duty should be strictly required, so that the track-way should be kept in order. The waste matter from the surface of the road and the gutters is valuable for manure, and if swept into heaps, would be carried away gladly for its worth by people living near, but in no case should it be thrown out upon the sides.

The slopes of the cuttings and embankments, ought to be neatly trimmed, and grassed over, and not be permitted to wear the ragged look now too common. The cost of keeping the borders of the roads in this way would be trifling, and would influence the residents upon them to keep their grounds, and dwellings in a better condition. A man will dress better, and use choicer language, when among gentlemen than when among clowns.

This system is intended to apply only to the highways or county roads. The care of the town roads, those established by the town for local convenience, may be left, as now, if thought best, in the hands of the town authorities.

County Commissioners.

The present method of laying out new roads, or altering the lines of the old ones by the county commissioners need not be changed, except that for a projected new road or alterations, the surveys, maps, profiles, estimates of cost, etc., shall be made under the direction of the chief of division, or resident engineer, of the county or district, and when ordered the new road shall be constructed, or the old one rebuilt under his charge. At present the county commissioners are liable to make mistakes in the location of roads, sometimes involving needless expense in their construction, and do so by establishing inferior routes, with steeper slopes than need be, or with greater deviations than are demanded by public interests or required by the nature of the ground. Not appreciating the importance of employing a competent engineer to ascertain the best route, and therefore not feeling justified in incurring a slightly greater expense, or being willing to gratify an importunate neighbor, they generally get a land surveyor, who knows nothing of road making, to mark out the line and to determine (rudely perhaps) the elevations along it, as well as calculate in a rough way the cubic contents of the cuts and fills, and reserve to themselves the duty of drawing up specifications which vaguely indicate the manner in which the work shall be done. It is next to impossible in this way, to avoid making mistakes. No one with an unpractised eye can select the best ground, where there is any choice, upon which to build a road, nor can any one with the use of instruments, without skill acquired from practice, fix upon the most feasible line without much unnecessary labor. The commissioners are usually men of other professions and share probably the mistaken estimation in which engineers are held by the public generally. The common impression that engineers are extravagant, and that they look only to the achievement of ends, without regard to expense, is exactly the reverse of the fact. Their great aim and study is economy ; economy in every form consistent with durability and adaptedness for use ; economy of money, material, time and labor. This may be said to be the secret of the profession. Without this economy very few of the great modern works would have been accomplished ; for it is only by applying labor and means to the very best advantage, and arranging costly materials in structures with scientific accuracy, so as to secure the greatest strength and durability, with the least possible expenditure of materials, that such works have been made practicable, or perhaps possible.

Should the system here proposed not meet at once with a favorable consideration, it is to be hoped that some different way of constructing new roads will be adopted. Let there be a resident engineer appointed for each county, who shall be employed by the commissioners to make the necessary surveys, &c., for projected roads, and superintend the construction according to a standard fixed by law.

If the people were as clamorous for good roads as for new ones, there would be less call for new ones, and the whole community would be benefited.

Another and final reason for the adoption of some system of responsible superintendence in the construction and care of roads, is that streams are often obstructed and their habits changed, by the improper location or construction of bridges. Much injury sometimes results from these causes. In flat lands, upon slow-flowing streams, large tracks are sometimes flooded more, and for longer periods, from a deficiency of water-way under a bridge, or some other mischievous interference with the natural current by the structure. In hilly regions, the bridges themselves are sometimes swept away and great injury is done to property on the borders of streams, from ignorance or carelessness in cramping a swollen torrent into too narrow a space by the misplacement of the abutments and piers of bridges, in order to shorten the span a few feet.

The drainage of lands, now held to be so important for agricultural and sanitary reasons, is very commonly injured or wholly prevented, from the want of proper culverts under roads.

The writer has often been told by Western farmers, living on lines of railroads, that they had at first been opposed to the roads crossing their lands as a serious damage, but that they had found that the drainage effected by the cuttings in the soil, more than compensated for the damage done by dividing up their fields. The building of a road, if properly managed, need at least be no injury to the drainage of lands, and in many cases should be a benefit.

Any person who has ever had experience in travelling upon our roads in their worst condition in the spring months of the year, will be thankful for any plan, within the public means, of making it comfortable; and when the people of the State become convinced that the evil of bad roads is not necessarily insurmountable, a remedy of some kind, more or less effective, will be applied. If it were possible to present the contrast between the present imperfect roads and well-made ones to actual view, no one would hesitate to accept any practicable method of insuring the well-made ones. Could the people have a trial, for a day, of perfectly good roads, they would never again do without them, if means and skill within reach could command them.

The plan proposed in this paper is believed to be the best that can be offered, and is submitted with confidence to the consideration of the people and law-makers of the Commonwealth.

The State of Massachusetts is usually foremost in the march of improvement, and it is to be hoped that she will head the advance in this new way of progress, which so plainly leads to an increase of prosperity and a still higher grade of civilization.

The work lies straight ahead; and with the leadership of His Excellency the governor, and the ready response of the legislature, there is a fair promise that we shall soon achieve one of the greatest and most needed of public improvements—a grand and complete system of really GOOD COMMON ROADS.

DEDHAM, January 28, 1870.

INDEX.

APPENDIX.

For the sake of aiding in arriving at some practical end, I have requested the gentlemen to whom the prizes for the preceding Essays were awarded, to suggest what form of legislation would be desirable as a change from our present inefficient system of road management, to one which should promise better, more economical and more satisfactory results. The large and varied experience and observation of these gentlemen, all of whom are competent engineers, entitle their opinions and judgment to favorable consideration; and the following, submitted by them, may serve as a basis or outline for future legislation. C. L. F.

An Act for the more perfect construction and maintenance of the common roads or highways throughout this Commonwealth.

Sect. 1. Establishes a body to be known as the state board of highways and bridges, to consist of three skilful civil engineers, or persons practically expert in the science of road making, to be appointed by the governor with the advice and consent of the council, and to have their office in the State House.

Sect. 2. It shall be the duty of the attorney-general, personally or by his deputy, to give his counsel and opinion on such matters as he may be called upon by the board, for which service his compensation shall be

Sect. 3. The first appointment of members of the board of highways and bridges shall be made on or before , and there shall be appointed one member each for the terms of one, two and three years; after that there shall on or before each year be appointed one member for the term of three years.

Sect. 4. Each member of the board shall receive an annual salary of dollars; give bonds for the faithful discharge of his duties; pay over all moneys, papers, &c., at the expiration of his term or when ordered by the governor and council.

Sect. 5. Board are to elect a president and treasurer, and make their own by-laws.

Sect. 6. A majority of the board constitutes a quorum; records to be kept of all the proceedings; copies of all plans, estimates, &c., to be kept; report to be rendered on or before each year, or when required by the governor and council. Each member authorized to administer legal oaths.

SECT. 7. Said board shall prepare and submit to the legislature a plan for the systematic classification of all the highways and townways in this Commonwealth into two or more of the following three classes:—

Class 1. State roads, to be controlled and maintained wholly by the State.

Class 2. District roads, to be controlled and maintained by the State, but the expense thereof to be borne by the towns and cities of the districts in which said road shall lie, and the State, in such proportions as said board shall apportion.

Class 3. Town roads, to be controlled and maintained as now provided by law.

The construction of new roads, of the three classes above specified, to be done as follows:—

Class 1. State roads, to be laid out and built by the State, through the Board of Highways and Bridges.

Class 2. District roads, to be laid out, &c., by the county commissioners, as now provided, but the board to have the final approval or disapproval of the proposed plans and profiles for said road, and also to have the charge and superintendence of their construction.

Parties aggrieved by the refusal or neglect of county commissioners to lay out a road, to have the right to appeal to the Board of Highways.

Class 3. Town roads, to be laid out and constructed as now provided by law.

SECT. 8. The paying of road taxes by labor is hereby abolished, and all road taxes are hereafter to be paid in cash.

SECT. 9. Board shall have the special charge and superintendence, subject to the laws and resolves of this Commonwealth, of all the highways and bridges, and the public works appertaining thereto, which are or shall be executed or maintained, wholly or in part by this Commonwealth. They shall also perform such other duties as may be required of them by the general court or the governor and council.

SECT. 10. Whenever any highway or bridge, or public work appertaining to these two, shall come partly within the province of this board, and partly within that of any other State board, already constituted, then such subject shall be discussed and decided upon in a joint convention or conventions, composed of equal numbers of this and the said other State board, and some member by them chosen as presiding officer.

SECT. 11. All applications or propositions for improvements or new works, of the kind specified in section nine as coming within the province of this Board of Highways and Bridges, and intended to be laid before the legislature, shall hereafter be first made to this board. Upon receiving such application, board shall investigate same, and if they find such work necessary and proper, shall thus report to the legislature, with an estimate of the expense thereof; if they do not approve of such application, they shall report the reasons for their disapproval.

The board may also, in like manner, recommend, whenever they think proper, any improvements of the kind above specified, though no application has been made therefor.

SECT. 12. It shall be the duty of the board to procure for the legislature full plans and estimates of contemplated works or improvements when so ordered by the legislature.

SECT. 13. Whenever any work shall have been authorized or ordered by the general court and the money appropriated therefor, board shall advertise for proposals for doing said work; plans and specifications of the same first to be placed on file in office of board, which plans and specifications shall be open to public inspection; advertisement to state work to be done and to be published ten (10) days at least. The bids shall be sealed bids, directed to board and accompanied by bond to the Commonwealth signed by bidder and two responsible sureties, in sum of two hundred ($200) dollars, conditioned he shall do the work if awarded to him, in case of his default to do so, forfeits, &c. Bids to be opened at time and place mentioned in advertisement.

SECT. 14. All contracts shall be awarded to the lowest responsible bidder and who sufficiently guarantees to do work under superintendence and to satisfaction of board; provided that the contract price does not exceed the estimate or such other sum as shall be satisfactory to board. Copies of contracts to be filed with state auditor.

SECT. 15. Board reserves right in contracts to decide questions as to proper performance of work and meaning of contracts; in case of improper construction may suspend work and relet the same; or order entire re-construction; or may relet to other contractors and settle for work done, &c. In cases where contractor properly does work, board may in their discretion as work progresses, grant to said contractors estimates of amount already earned, reserving fifteen per cent. therefrom, which shall entitle holder to receive amount, all other conditions being satisfied.

SECT. 16. In case prosecution of any public work be suspended, or bid be deemed excessive, or bidders be not responsible, board may, with written approval of governor, where the urgency of the case, or interests of the Commonwealth require it, employ workmen to perform or complete any work ordered by the legislature: provided, that the cost and expense shall in no case exceed the amount appropriated for the same.

SECT. 17. All supplies of materials, &c., when costing over five hundred ($500) dollars, to be purchased by contract, subject to same conditions as letting out work.

SECT. 18. Whenever board think necessary, for interests of the Commonwealth, to protect same from damage or loss, shall report thus to governor and council and reasons for same, asking power to give contracts without notice required above, and governor and council may grant request, provided three-fourths vote for it.

SECT. 19. Whenever board is of opinion a work may be done better without a contract, shall so report to legislature, and they shall procure machinery, materials, &c., hire workmen, &c., to do said work, whenever so authorized by the legislature.

SECT. 20. All contracts and bonds by board to be in the name of the Commonwealth.

SECT. 21. No member of the board to be interested in any contract; all contracts made with any member interested, governor may declare void, and shall remove such member so interested from office. It is the duty of every member of the board and every officer of the Commonwealth to report any such delinquency, if discovered.

SECT. 22. Board shall be empowered to employ such engineers, clerks or other assistants, as shall be provided for by the legislature.

SECT. 23. Board shall, on or before every year, submit to the auditor, by him to be presented to the legislature with his annual estimate, a statement of the repairs and new work needed for the current year, and of the sums required by the board therefor; report to be in detail; all sums appropriated therefor to be included in the annual tax-levy.

SECT. 24. All moneys to be paid to any person out of moneys so raised, shall be certified by president of board to auditor, who shall draw warrant on treasurer therefor, stating to whom payable and to what fund chargeable; such warrant to be countersigned by president of board.

SECT. 25. Board to keep accounts, showing moneys received and spent, clearly and distinctly, and for what purpose. Accounts to be always open for inspection of auditor or any committee appointed by the legislature.

www.ingramcontent.com/pod-product-compliance
Lightning Source LLC
LaVergne TN
LVHW021422110826
845150LV00007B/2035

* 9 7 8 1 4 2 5 5 0 8 9 1 3 *